ELEMENTS OF GEOGRAPHICAL HYDROLOGY

Brian Knapp

*Leighton Park School,
Reading*

London
UNWIN HYMAN
Boston   Sydney   Wellington

Published by the Academic Division of
Unwin Hyman Ltd
15/17 Broadwick Street, London W1V 1FP, UK

Unwin Hyman Inc.,
8 Winchester Place, Winchester, Mass. 01890, USA

Allen & Unwin (Australia) Ltd,
8 Napier Street, North Sydney, NSW 2060, Australia

Allen & Unwin (New Zealand) Ltd
in association with the Port Nicholson Press Ltd,
Compusales Building, 75 Ghuznee Street, Wellington 1, New Zealand

First published in 1979
Third impression 1989

Transferred to Digital Printing 2004

British Library Cataloguing in Publication Data

Knapp, Brian John
    Elements of geographical hydrology
1. Hydrology
I. Title
551.4′8   GB662.3   78–40373

ISBN 0-04-551030-X

Typeset in 10 on 11 point Times

CONTENTS

TABLES

PREFACE

Although the science of hydrology has held a firm place in higher education and industry for many years, it has only recently found a formal place in the 'A' level syllabuses. With the upsurge of interest in water resources in particular, some aspects of hydrology have become more widely known. Nevertheless the central role that hydrology plays in a number of subjects, such as geomorphology and pedology, has been less effectively publicised. As a result, the purpose of this book is not only to introduce the theory of hydrology, but also to demonstrate its relevance in the real world by relying on detailed and wide-ranging specific examples.

The book has been planned in the hope that it will be available to students for the whole of their 'A' level courses. As such the first two chapters provide, as well as hydrological theory, an adjunct to a more conventional geomorphology text. Chapters 3, 4 and 5 build on the foundation of the early part of the book but are largely independent of one another and can be used in conjunction with books on pedology, slope formation and resource development. For these later chapters it is expected that the student will already have some basic knowledge and will use the material included in this book to give a hydrological slant to their further studies.

To help students consolidate their understanding of hydrology, there are selected problems at the end of most chapters. However, these problems are not repetitions of examples in the chapters and are to be seen as integral advances in each topic. The nature of the data is such that it may be analysed at a variety of levels. Students taking courses at more advanced levels can use the examples with the confidence that they represent 'the state of the art' in each topic.

I have been particularly conscious of the need to provide up-to-date examples wherever possible and to reflect current thoughts in the subject. As in many rapidly advancing disciplines, not everyone will agree on each point of detail and the responsibility for particular viewpoints, omissions and the arrangement of the work is my own.

B.J.K.
Woodley 1978

ACKNOWLEDGEMENTS

A work of this kind involves the active cooperation of a large number of people in the supply of data and to them I would like to tender my thanks. However, I am particularly grateful to Denys Brunsden, Richard Chorley, Ian Fenwick, Roger Jones, Mike Kirkby, Pam Wilson, Peter Worsley; the surface hydrology section of the Thames Water Authority, especially Stewart Child; the Library staff of The Central Water Planning Unit; and the Institute of Hydrology at Plynlimon for their advice and encouragement. Finally I would like to thank the following for help in critically reviewing the typescript: Frank Button, Ray Jessop, Christopher Rogers, Darrell Weyman, Graham Agnew, Mrs M. Walls, Mr K. Briggs, Richard Huggett, John Rolfe, Mr J. A. Williamson and Nigel Bates.

I would like to join with the publishers in acknowledging permission given by the following to reproduce copyright material:
Edward Arnold for Figure 1.17; Institute of Hydrology for Figures 2.2, 2.7; Canadian IHD Committee for Figures 2.11, 2.12; Norges Vassdrags og Elektrisitetsvesen for Figure 2.15; P. Worsley for Figures 2.13, 4.1; *Int. Assn Sci. Hydrol. Bull.* for Table 4.2, Figure 2.18; US Geological Survey for Figures 2.20, 2.21, 2.29; Institute of British Geographers for Figures 2.25, 4.13, 4.14, Table 4.3; Bundesanstalt für Gewasserkunde for Figures 2.28, 5.15; Elsevier Publishing Company for Figures 1.12, 2.5, 2.32, 2.33; I. M. Fenwick for Figures 3.4, 3.7; Soil Survey of England and Wales for Figures 3.9, 3.26; Soil Survey of Scotland for Figure 3.11; C. W. Mitchell for Figure 3.22; Cambridge U.P. for Figures 4.2, 4.3; Institution of Civil Engineers for Figures 4.5, 4.7; *Geografiska Annaller* for Figure 4.9, Table 4.1; Arctic and Alpine Research for Figure 4.10; Central Water Plannning Unit for Figure 5.3; Institution of Water Engineers for Figure 5.5; Severn–Trent Water Authority for Figure 5.8, Tables 5.2, 5.3; V. Sagua for Figure 5.9; Water Pollution Control for Figures 5.13, 5.14; Crown copyright is reserved on Figures 3.8, 5.17, Tables 5.1, 5.5; Water Data Unit for Table 5.6; National Water Council for the extract on p. 78; Loyd Martin for the data in Example Problem (D) on page 40; US Department of Agriculture for Figure 2.3; Elsevier Publishing Company and the Institute of Hydrology for Table 1.1.

PART ONE THEORY

Chapter 1

PROCESSES

Introduction

More than two-thirds of the Earth's surface is covered by water, yet less than 3% of it is available on land at any one time to provide the means of landscape evolution and support life. It has long been clear that the water which flows down rivers to the sea must somehow return to the land for these processes to continue and from this has evolved the concept of the **hydrological cycle**.

We can begin with the oceans, the great storage elements of the cycle. Evaporation from them provides the moisture for cloud formation and hence precipitation. Then, from the moment water reaches the land as rain to the moment it returns to the sea, it is engaged in a sort of obstacle course in which only a certain percentage reaches its goal. Much is returned direct to the atmosphere by evaporation, while some is held in various temporary stores such as soil or rock. Because of this it is often convenient to consider the land-based part of the hydrological cycle as a linked series of cascading reservoirs, each with a limited storage capacity (Fig. 1.1). It is the role of the hydrologist to attempt to understand each part of the hydrological cycle, but with an emphasis on those parts that are most directly connected with the land.

Precipitation

The land-based part of the hydrological cycle begins with a study of **precipitation** which, although largely within the province of meteorology and climatology, is introduced here in order to explain many runoff processes.

All precipitation results from atmospheric cooling and subsequent condensation of water vapour. Usually such cooling happens either at the frontal zones of depressions, or because of convection, or because an air mass is forced to rise over an area of high land. In addition the amount and distribution of precipitation depends on the global atmospheric circulation. Some regions, such as the Sahara and Antarctica, experience very little precipitation as they are permanently under the influence of the high pressure zones of the global circulation (Fig. 1.2). Continental interiors such as the south-west of the USA, and central Europe, which are away from the main areas of high pressure, often become very hot in summer and produce conditions just right for convectional instability. The resultant storms are important hydrologically because of their very high intensity, localised occurrence and short duration. More persistent areas of instability occur under the influence of the

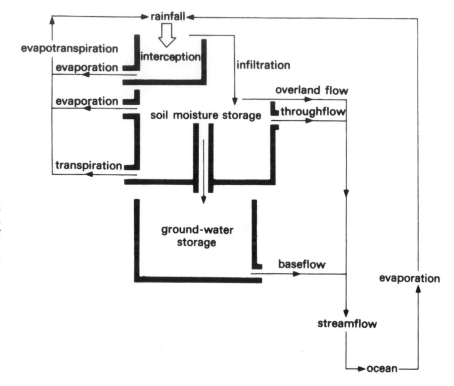

Figure 1.1 The land-based part of the hydrological cycle can be thought of as a series of storage units. As each store fills after rainfall, water cascades to the next unit, eventually reaching the sea.

Figure 1.2 The general global circulation determines where rain will fall and its dominant form.

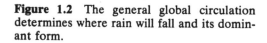

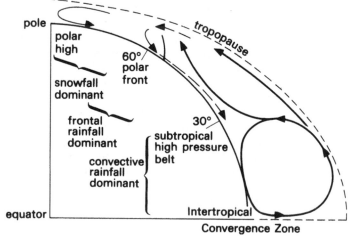

Intertropical Convergence Zone (ITCZ) and convective storms become a daily phenomenon.

In the middle latitudes precipitation mainly results from the passage of fronts. These provide a much more uniform regime than do convectional storms, with precipitation generally of low intensity and frequent occurrence. Nevertheless within frontal systems there are usually convective cells giving higher precipitation intensities and it is such variations that often make frontal storms so complex. Further precipitation in such regions is provided by small convective showers which form in unstable polar airstreams.

The main effect of upland is to force an air mass upwards thereby promoting cooling and condensation, even without frontal or convective uplift, which increases the frequency of precipitation.

But whether precipitation is caused by fronts, convection or relief, there is often a marked seasonal pattern in the regime and this gives river flow a corresponding seasonality. Such effects are primarily caused by a shift of atmospheric circulation patterns on a global scale with the attitude of the sun as it moves between the Tropics. Such seasonality is particularly pronounced in tropical areas where the 'wet' season corresponds to the influence of the ITCZ and the 'dry' season to the subtropical high pressure belt.

Ice crystals (snowflakes) fall from most clouds in middle and higher latitudes even in summer. However the form they take by the time they reach the ground is determined by the temperature of the air through which they fall. Thus in summer, high air temperatures cause melting and rain falls, but in winter there is a greater likelihood of snow, especially in continental interiors where temperatures are very low.

It is important to understand that there is a fundamental hydrological difference between snowfall and all other forms of precipitation. This is because snow usually remains on the land surface for a long time after it has fallen; it frequently drifts; and it is only available to the hydrological

Figure 1.3 Rainfall interception on leaves.

cycle during spring melting. By contrast all other types of precipitation are more or less immediately available. Rainfall is measured in millimetres depth of accumulation in a specified time period. Snowfall is converted to rainfall equivalent.

Interception

When precipitation reaches the ground, it is **intercepted** by a great variety of dry surfaces which need to be thoroughly wetted before they will transmit water (Fig. 1.3). The importance of interception varies considerably depending on factors such as surface roughness. In the case of a forest cover, there are also many levels of surface to be wetted because drips from the upper wetted leaves will largely be intercepted in turn by lower leaves (Table 1.1). This is, of course, why rainfall takes such a long time to reach the ground beneath trees. By contrast in urban areas there is much less loss as the surfaces are smooth (roofs, roads, etc.). However in all circumstances it is the first part of the precipitation that will be lost to interception and this is a fixed amount for the surface concerned. After this initial loss all further precipitation will

Table 1.1. Interception and evapotranspiration losses for a mature spruce forest, Thetford, Norfolk (1975)

4 week period commencing	Gross rain (mm)	Rain falling through forest canopy (mm)	Rain running down tree trunks (mm)	Interception by bracken (mm)	Total interception (mm)	Calculated transpiration (mm)	Total evapotranspiration (mm)	Moisture balance (mm)
1–1–75	39·1	22·6	0·9	nil	15·6	4·5	20·1	+19·0
14–7–75	38·3	22·5	0·1	5·4	21·1	68·5	89·6	−50·5

be shed so that it is relatively insignificant in a long storm but very important in a short shower. Again, because of high evaporation rates in summer, surfaces dry very quickly so that there is a full interception loss for each new shower. In winter, with lower evaporation rates, surfaces dry only slowly and a full interception loss may not occur with each new precipitation event. Interception loss is measured in millimetres per storm event.

Evaporation and transpiration

Whenever unsaturated air comes into contact with a wet surface, a diffusion (or sharing) process operates. This process, **evaporation,** happens at the surfaces of lakes, rivers, wet roads, and even raindrops as they fall from clouds. When the water body is not large, as in the case of water intercepted by road surfaces, it may be evaporated completely and quickly, leaving the air still unsaturated. Precipitation leaving the cloud base also loses mass but it usually falls too quickly to be entirely evaporated. Evaporation is measured in millimetres per time period in the same way as precipitation (e.g. mm/h).

The rate of evaporation is dependent on several factors, the most important being a source of energy for vaporisation. This is largely supplied by solar radiation which is at a maximum in summer, making evaporation take place more quickly in summer than in winter. In addition, warm air will hold more moisture than cold air, and dry air will take up moisture faster than wet air. However, if air remains completely still over a wet surface, the air soon becomes saturated and no further evaporation takes place. Wind is therefore needed to bring fresh, unsaturated air into contact with the wet surface.

Evaporation takes place continually from a lake or a river and this is the maximum continual loss of water possible – the **potential evaporation.** However in most cases the surface area covered by water is small and most evaporation takes place from soil and plant surfaces. Water is not easily transferred from the soil body to the surface, so the surface quickly dries out. After this, although some evaporative loss does occur, it is well below the potential rate and is of minimal significance (except in arid environments).

In most humid regions the proportion of the land surface subject only to evaporation is small; for most of the year soil surfaces are covered by vegetation and in these circumstances **transpiration** will be the major factor determining total losses of water from the land surface. Transpiration is the process whereby water vapour escapes from living plants mainly by way of leaves. This loss of water through plant surfaces is replenished by water being drawn up through the plant tissues from the root hairs which are in contact with soil water. Transpiration is a very powerful process because it draws upon the reserves of water stored in the soil pores at depth and is in direct contrast to evaporation which simply dries out the soil surface.

The factors controlling the rate of transpiration are similar to those for evaporation. In addition, however, there are controls exerted by the soil and the plant itself. There is still no general agreement about the exact nature of the factors controlling plant transpiration, but it seems that water can move freely to leaves only if there is an unrestricted supply available to the roots. This does not mean that the soil has to be saturated, but clearly as moisture is lost from soil pores fewer roots will be in contact with water and supply to the plant will be restricted. Transpiration is really a 'leakage' from leaves over which many plants have very little control. As a result it continues at near the potential maximum rate until a stage of soil moisture deficiency is reached when the uptake of water no longer balances transpiration. After this the difference has to be obtained from within the plant itself and this causes wilting. The **wilting point** corresponds to a very low soil moisture value which is not often reached in humid regions except on very shallow or permeable soils. However in arid and semi-arid regions the wilting point would regularly be reached in plants unadapted to the prevailing conditions and so indigenous plants have special leaf configurations (e.g. cacti). Special leaf forms and the general scarcity of vegetation make the role of transpiration very much less important in arid and semi-arid regions.

Clearly it is very difficult to separate evaporation from transpiration, so the term **evapotranspiration** is used to describe the combined effect of the two as they influence storage of precipitation in the soil. This total loss by evapotranspiration is of major importance and may account for about three-quarters of all precipitation in a year, even in humid regions.

The role of evapotranspiration in the terrestrial part of the hydrological cycle may thus be summarised as the progressive loss of soil moisture to a degree far beyond that which would occur through gravity drainage alone. A moisture deficit is thus created which has to be replenished before any fresh rain water is available for the runoff process. This replenishment takes time and is partly responsible for the lag between rainfall and runoff. Finally it is necessary to stress that the importance of evapotranspiration is dependent on a number of climatic factors and so, although clearly at a maximum in summer, losses are very variable (Table 1.1).

Infiltration, throughflow and overland flow

When precipitation has completely wetted any vegetation, the remainder is available to wet the soil or to run off over the surface (**overland flow**). The process whereby water enters the soil is called **infiltration** and it is normally measured in centimetres depth of water per time period.

Soil acts as a sort of giant sponge containing a labyrinth of passages and caverns of various sizes. The total space available for water and air is called **porosity** and is expressed as a percentage of the soil volume. Some of the passages and caverns (**soil pores**) are interlinked, although others are cul-de-sacs and conduct no water. The passage of water from soil surface to stream bank or water-conducting rock (**aquifer**) thus takes place in a very tortuous way through small-diameter passages, so it is not surprising that water movement is slow. The speed of conduction of water is determined by the proportion and size of the interlinked water-filled pores. If all pores are full, the soil is said to be **saturated** and the plane defining the upper surface of the saturated zone is called the **water table**.

Figure 1.4 shows the soil saturated for a depth H above a plane XY. Under these conditions the speed of water movement varies directly with H (the **hydraulic head**) and is related to it by a constant called the **permeability** (hydraulic conductivity), which is measured in centimetres per second. Sandy soils have, on average, larger pores than clay soils and hence a higher conductivity value. When soils are unsaturated, water still flows, but in fewer pores. In this case the flow rate also depends on the ratio of pores still containing water to those containing air (the **soil moisture content**) and it is very much lower than the saturated rate. Moisture content is often measured as a percentage relative to the water-holding capacity at saturation.

Most soils are on sloping land and so soil-water movement takes place towards the slope base. The vertical component of this movement is called **percolation**, the downslope component **throughflow** (Fig. 1.5), both are normally measured in centimetres per second. Unless the soil is very dry or there is a storm of great intensity, all the effective rainfall will infiltrate the soil, filling up empty soil pores and establishing a zone of higher water content near the surface. The leading edge of the newly penetrating water is called the **wetting front** and it separates a zone of high permeability above from one with drier soil and a lower permeability below. If the storm is sufficiently prolonged, the wetting front eventually reaches the soil base and all storage is filled. Thereafter water moves downslope by throughflow, except in the special case of the part of the soil adjacent to a stream. Here,

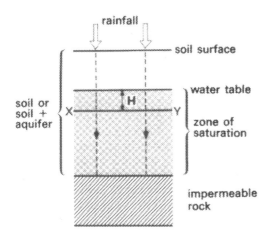

Figure 1.4 The position of a water table.

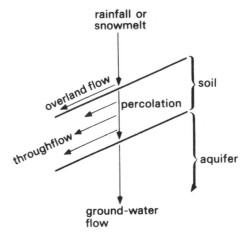

Figure 1.5 Water flow on a hillside.

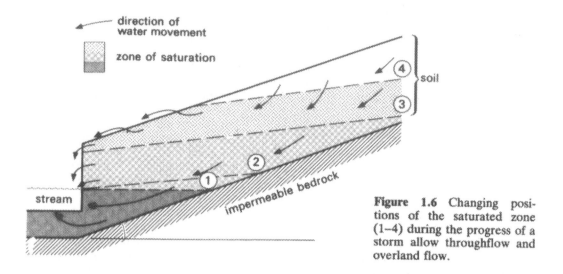

direction of water movement

zone of saturation

soil

4

3

2

1

stream

impermeable bedrock

Figure 1.6 Changing positions of the saturated zone (1–4) during the progress of a storm allow throughflow and overland flow.

because the soil is in contact with the stream, it is saturated (Fig. 1.6 (1)) but water will only flow out of the soil when the water table rises above the stream level creating a **hydraulic gradient**. Percolation and throughflow cause a build up of water in the lower part of the soil, elevating the water table (Fig. 1.6 (2)). During a storm a large amount of water reaches the slope base as throughflow, causing a build up of the saturated zone (Fig. 1.6 (3)). This in turn causes a greater throughput of water to the stream. In some cases the amount of expansion of the saturated zone will be great enough for the water table in the soil to intersect the surface. When this happens, water will seep out of the soil surface as a very shallow sheet of water known as overland flow (Fig. 1.6 (4)). Thus with the progress of a storm a saturated zone may build up in the soil adjacent to the stream network sufficient to allow overland flow to occur. Overland flow is rapid and can be compared with the rate of flow in shallow channels. Any rain falling on saturated soil will also contribute as a component of overland flow to the supply of water to the stream channel. If a storm is of low intensity or short duration, there will be insufficient throughflow to enable the saturated zone to rise to the ground surface. As a result all contributions to stormflow (except that due to direct channel precipitation) will be from throughflow. If on the other hand the storm continues, and especially if the intensity increases, the saturated zone will

build up, overland flow will occur and with it larger and more rapid contributions will be available to the stream. The first part of both contributions will be the same, the divergence occurring when overland flow takes place. Clearly the longer the storm and the greater the total rainfall, the larger the saturated zone will be as will the area over which overland flow occurs. This **contributing area** expands and contracts with the size of the storm but is nearly always adjacent to the stream network. Peak flow is thus controlled by the nature of the contributing area only (Fig. 1.7).

The role of headwaters, and indeed any hillside **concavity** (hollow), in producing rapid response to rainfall will be clear from Figure 1.8. Water flowing down the hillsides follows the line of steepest gradient; that is, at right angles to the contours. As a result, unit width of hillside soil x has water from width x' directed into it by throughflow, thereby increasing the water content of soil in the concavity. It becomes very likely therefore that saturated conditions will prevail within a short period of the onset of rainfall in such localities. After saturation has been reached, any further concentration of water into the concavities can be accommodated only by overland flow and this can occur at rainfall intensities well below those needed to exceed infiltration capacity, as is the case for regions bordering streams. The result in this case is a concentration of overland flow into point sources called **seepages** or **springs** depending upon

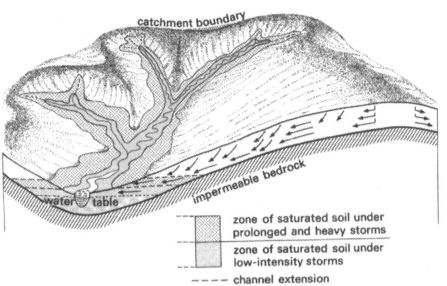

Figure 1.7 The effect of a rainstorm on a catchment is to cause stream extension and expansion of the saturated zone adjacent to stream channels. Note that the soil thickness has been exaggerated for clarity.

the vigour of the source. They are established above the normal stream channel for a short time during and after the storm rainfall, their concentrated flow naturally running into the established stream network and causing stream extension. It is because of the direct link to streams that these headwater seepage zones are so important in helping to generate stormflow peaks.

It seems that stream channels occupy between 1 and 5% of a river basin (**catchment**) area in humid regions and that the contributing area extends up to about 30% of the catchment in a prolonged storm. Water moving in unsaturated soil outside this region moves much more slowly than in the saturated region, but even here saturated velocities will normally be less than 50 cm/h. Thus the peak of the stream hydrograph is often passed before the majority of the slopes have even filled their moisture deficits.

Soon after the rainstorm ceases, streamflow **recession** begins and this again focuses attention upon the properties of the soil. The same soil pores that prevented water from reaching the stream quickly are now effective in maintaining a store which keeps the streams flowing between rainstorms.

About 48 hours after rainfall has ceased, gravity will have drained the larger pores, resulting in a soil moisture content often called **field capacity**. When these have been drained, only smaller pores are available and the conducting routes are therefore more tortuous. This is why soils drain progressively more slowly over time, often at a

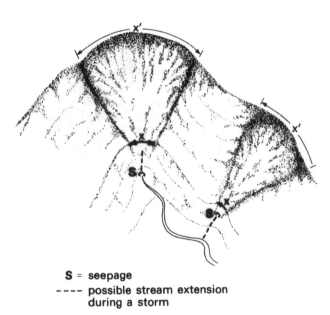

S = seepage
- - - - possible stream extension during a storm

Figure 1.8 The presence of hollows and concavities serves to concentrate flow and promote soil instability.

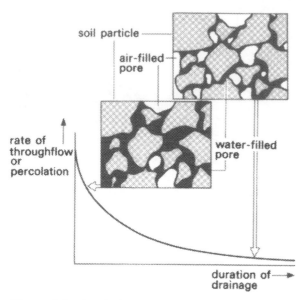

soil particle
air-filled pore
rate of throughflow or percolation
water-filled pore
duration of drainage

Figure 1.9 As drainage proceeds, fewer pores are water-filled and permeability decreases.

near-exponential rate (Fig. 1.9). It is of course such unsaturated throughflow that eventually finds its way to the saturated zone adjacent to the stream and provides part of the base flow. In general the deeper the soil, the greater the storage potential before the water table rises to the surface and allows overland flow. As a corollary, shallow soils have a low storage potential and so promote overland flow.

Infiltration depends very much on the behaviour of water in the soil, because the permeability has to be greater than the rainfall intensity. However in some cases, such as are common in semi-arid and continental regions, rainfall intensity may be just too great for the soil pores to transmit water fast enough. In these cases the infiltration rate will be exceeded irrespective of texture, even at the start of the storm when storage capacity is still large. The sort of overland flow that results can occur on any part of the slope and not only on land near a stream. In such cases the probability of soil erosion is high, as is typified by the gullying of slopes in south-west USA.

It should also be remembered that infiltration is dependent on maintaining surface pores, so that pressure of machinery and trampling by animals can compress the surface pores. Even raindrop impact on bare soil can break up soil structural units and wash fine particles into pores, so reducing their size.

Ground-water flow

In some cases catchments are underlain at least in part by aquifers. These rocks (usually chalk, limestone or sandstone) contain a large number of fissures which provide the main passageways for water movement. In some cases (e.g. chalk) these fissures may be very small and so behave in the same way as pores in a soil; in others (e.g. Carboniferous limestone) there are large fissures separating massive blocks that behave in the same way as a system of sewer pipes. The speed of water movement and capacity for storage vary considerably depending on the rock type, but they do provide catchments with an enlarged storage element.

When a soil is underlain by permeable rock, there is no accumulation of water at the soil/rock junction except in very high flow conditions, so percolation continues through the soil and into the rock. Soils on permeable rocks will therefore have a very small throughflow component; they will become saturated less readily and overland flow and rapid catchment response to rainfall will be largely eliminated.

Water moving in an **unconfined aquifer** (a permeable rock with no impermeable stratum above) slowly sinks to its lowest possible level so that at depth all fissures are water filled and saturated conditions prevail. As with soils, if the water table intersects the ground surface, springs will be produced and water will flow out. Springs in chalk and sandstone respond very slowly to new rainfall events because of the slow transmission of pressure following a rise in the water table (Fig. 1.10). The delay is caused by the small size of the fissures which restrict the flow of water, and is in marked contrast to the massive limestones like the Carboniferous formation. Here large impermeable blocks are separated by fissures, much widened by solution and abrasion through which water is transmitted at a very much higher rate than in chalk or sandstone (Fig. 1.11). This of course also means that the height of the water table varies much more rapidly in limestone strata, and that they drain more quickly and have a fairly low storage potential. Deep throughflow and groundwater flow together make up **base flow**, normally measured in cm^3/s.

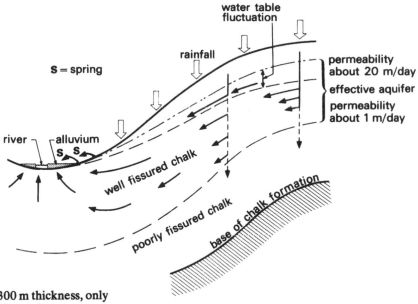

Figure 1.10 In a chalk stratum of 300 m thickness, only the upper 50–60 m is sufficiently fissured to act as an aquifer. In this upper zone, outflow to streams following a storm results from an increase in hydrostatic pressure.

Figure 1.11 Water flow through passages in a massive limestone stratum. Contrast this with Figure 1.9.

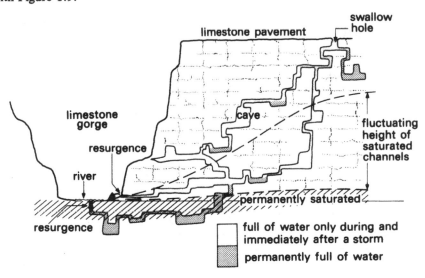

Streamflow

Streamflow is one of the easiest parts of the hydrological cycle to measure, yet it represents the complex contributions from all the sources so far described. Streamflow discharge is normally measured in cubic metres per second (**cumecs**). A plot of the measured flow against time at any section of a channel is called a **hydrograph** (Fig. 1.12). If the storm continued at a steady rate indefinitely, flow out of a catchment would reach and maintain a steady high value. For example, 1 mm rainfall per hour would yield about 30 cumecs from a 100 km² catchment. However, storms do not last long and have usually stopped before the steady value is reached; therefore hydrographs show a peak, followed by a recession limb which is often almost exponential in form, declining to a progressively steadier low value if no further storms occur. Figure 1.12 represents a storm of sufficient size to make overland flow a major contributor, but when this is not so, the hydrograph is more like that of Figure 1.13.

If a storm falls uniformly over a catchment, all hillsides will contribute to channel flow at the same time (Fig. 1.14). However there will still be a lag before most of the water reaches the measurement point because of the channel lengths involved. The peak flow comes when water from the maximum number of sources (i.e. **A**, **B** and **C**) are contributing to the outflow at the same time. Initially only reach (**B**) will contribute because there has not been time for other reaches to provide water. A little later not only is (**B**) contributing, but water from reach (**C**) has now had time to travel distance

(L_3) and so augment the flow. Later contributions are also received from (**A**) and so on. Flow contributions decrease in the reverse order following a decrease in rainfall intensity.

These time lags are extremely important in determining catchment response and they are related to catchment size, shape, drainage density and average channel gradient. For instance Figure 1.15 shows the hydrograph resulting from a storm of uniform intensity over a catchment. Because of the longer channel length in **A**, the hydrograph is much more subdued than for **B**. A storm moving across the catchment can show similar effects on the hydrograph. In Figure 1.16 a storm moves from mouth to source to give a multiple-peaked hydrograph, whereas the same storm moving from source to mouth gives a single, much more sharply peaked hydrograph.

The analysis of streamflow data

It is often difficult to make quantitative comparisons between two catchments because of their size or different length of storm duration. A simple and direct method of hydrograph standardisation is to derive a **unit response graph** (unit hydrograph). This is the hydrograph of storm runoff (without base flow) that is produced by a storm of known effective rainfall. Figure 1.17 indicates the way such a hydrograph is produced. The steps involved are:

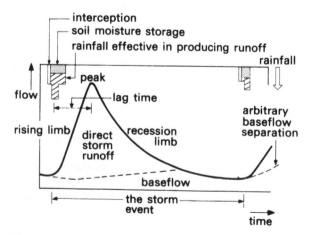

Figure 1.12 The characteristics of a storm hydrograph.

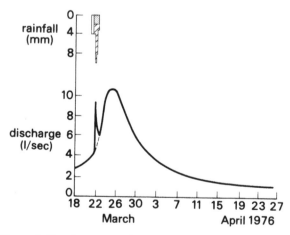

Figure 1.13 A storm on 22 March 1976 occurred over a small catchment in the Quantock Hills, Somerset and caused an immediate peak due to channel precipitation. There was no overland flow and the main hydrograph peak was produced by throughflow.

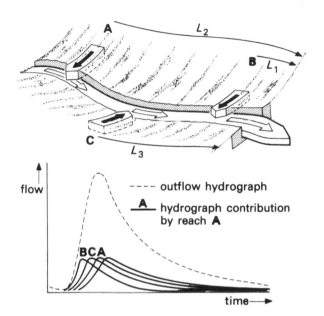

Figure 1.14   Lag within a catchment is partly caused by the time needed for water to travel within the stream network. Here throughflow contributions at **A**, **B** and **C** enter the stream via the banks at the same time but **B** is measured before **C** which, in turn, is measured before **A** because of the different stream lengths L_1, L_2, L_3. The hydrograph peak occurs when the maximum number of such contributions coincide at the measurement point.

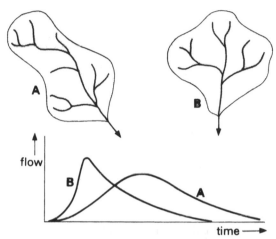

Figure 1.15   These two catchments have the same area and drainage density, but the elongated shape of **A** gives a longer channel length. Water near the headwaters is lagged behind that near the measurement point because of the time taken for movement along the channel. By the time it arrives, water from near the outflow point has already largely been through the measurement section. The result is a broad hydrograph without a sharp peak. The drainage network of catchment **B** is such that contributions from all parts reach the measurement point at nearly the same time and give a sharp hydrograph peak.

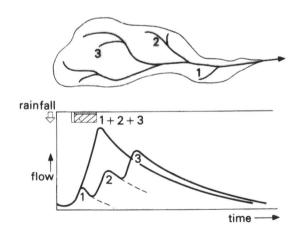

Figure 1.16   The way a catchment responds to rainfall is not just dependent on the physical characteristics of the catchment. As catchments get larger, a rainstorm is less likely to be of uniform intensity over the whole area and throughout its duration. In this example a storm of uniform intensity has moved from source to mouth in such a way that, despite the elongated shape of the catchment, all contributions arrive at the outlet together, giving a very sharp peak. The same storm moving from mouth to source will often allow the main runoff from the lower catchment to reach the outflow point even before the upper catchment has begun to contribute, thus producing a very complex hydrograph whose peaks represent the individual subcatchment characteristics 1, 2 and 3.

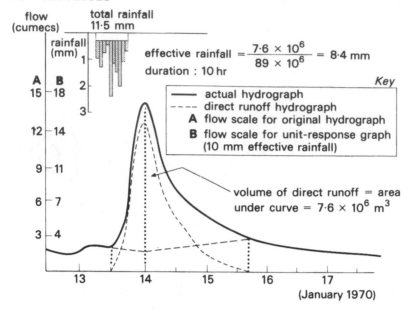

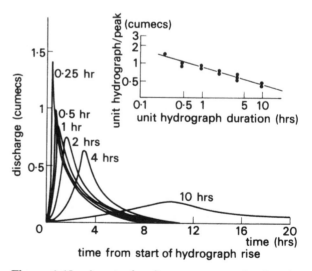

Figure 1.17 A six-hour unit response graph constructed for the River Mole, Horley, Surrey. The catchment area is 89 km², the total storm rainfall 11·5 mm.

1. Separate the direct runoff from the base flow by drawing an arbitrary base flow separation line. The exact slope of this line does not matter provided the same line is used in all analyses.
2. Draw a hydrograph of direct runoff only by subtracting the base flow component from the total hydrograph.
3. Measure the area under the direct runoff hydrograph by counting squares. This value is the volume of rainfall effective in producing direct runoff.
4. Divide this volume by the area of the catchment to give the depth of effective rainfall.
5. Throughout the storm, rainfall is used to replenish soil moisture deficits as discussed previously, it is only the surplus from areas near to the stream network that contributes to direct runoff. The rainfall graph can be divided into two sections using the value obtained in (3). The number of hours of effective rainfall is the storm duration (x).
6. To standardise the streamflow values to represent the effect of one centimetre (unit) of rainfall, divide each value by the effective rainfall in centimetres. The result is a response graph of one centimetre of effective rainfall of duration x hours.

Several examples are needed for the same storm duration to get a generalised response graph. A set of such response graphs illustrating the behaviour

Figure 1.18 A set of unit response graphs for the Rosebarn catchment (2·6 km²) near Exeter, Devon.

of a catchment for storms of varying duration can be built up as in Figure 1.18. Other response graphs can be sketched in or special techniques used to calculate them. Catchments of different areas can also be compared by the further refinement of: (1) dividing the streamflow values by the peak discharge; and (2) dividing the time units by the lag time.

Chapter 2

CATCHMENT SYSTEMS

Introduction

Chapter 1 examined the hydrological cycle in some detail to show how each part of it functions. This chapter brings the parts together and examines how they interact in real situations. Several small catchments are examined to illustrate the range of precipitation–streamflow patterns possible, and towards the end an attempt is made to inquire into how larger catchments represent the combined responses of many smaller ones.

Upland catchments

Upland catchments are characterised by steep slopes, well-defined boundaries, thin soils, high rainfall and low evapotranspiration. In these areas hydraulic gradients are steep, and streams respond rapidly to storms. The frequent and often heavy rainfall gives soils a low moisture deficit, so a relatively large proportion of the rainfall reaches the stream network. Streams have well-defined channels with an efficient cross-section which helps to transmit water effectively, whilst in many cases there is no flood plain to store water as the hillside slope continues right down to the stream bank (Fig. 2.1).

Low temperatures and low evapotranspiration tend to inhibit the growth of some forms of vegetation and in many cases peat develops beside the stream channels. Elsewhere coarse grass or tree plantations (especially conifer) are common. However, despite the high rainfall, overland flow is still rare except near the stream or in concavities in the hillsides (p. 10), so that throughflow dominates over much of the catchment. Most upland catchments are underlain by impermeable materials giving a negligible ground-water flow contribution. Consequently the recession limb of the hydrograph has to be sustained by soil throughflow (itself a limited resource) and so a relatively steep recession is to be expected.

Actual response times and magnitude of flood peaks following a period of rainfall will vary with: (a) the duration and intensity of rainfall; (b) the moisture deficit that needs to be replenished in the soil; and (c) the shape, size and average slope angle of the catchment. The way in which an upland catchment responds to rainfall is illustrated by the headwaters of the River Wye, Plynlimon, central Wales, an area of 10·5 km² with thin soils developed in impermeable shales and grits (Fig. 2.2). Soil texture is mainly silty clay, but a well-developed system of cracks between structural units (p. 48) helps increase the effective permeability. Occasionally these cracks have even become enlarged to the size of pipes. Figure 2.2 shows that a series of storms occurred on 5 August and these thoroughly wetted all the catchment and replenished the moisture deficit. The main storm rainfall fell late on 5 August and this resulted in a peak of streamflow between 0100 and 0300 hours on 6 August. Overland flow was observed from all the areas adjacent to the stream network as predicted above. However, most of the rest of the valley side slopes did not become saturated up to the surface, and throughflow dominated. A few concave areas did become saturated to the surface and these then formed an extension of the stream network. It is important to note that, even in this fairly extreme case, when at peak flow the catchment was contributing 5·4 cumecs/km², overland flow was still mainly confined to areas near the streams. The time lag between peak rainfall intensity and peak runoff was only two hours.

The Wye catchment has a vegetation cover of coarse grass, but the adjacent headwaters of the River Severn are more than half forested. In Chapter 1 it was suggested that vegetation cover might play a significant role in determining hydrograph response and this can be tested by comparing the hydrographs of the Severn and Wye for the same storm. The Severn catchment is slightly smaller

(b)

Figure 2.1 Part of the Wye Catchment, Plynlimon, central Wales. Here a tributary joins the main river which flows from right to left. (a) dry weather flow; (b) during a storm.

Figure 2.2 The effect of the storm of 5–6 August 1974 on the upper Wye and Severn as shown by unit response graphs. *l* is the time lag between peaks. Note a hillside profile and soil sequence is given in Figure 4.3.

than the Wye (8·7 km²), the average distance to stream channel is less and the average slope angle slightly greater, all factors that should lead to a more rapid response to rainfall. In addition there are many artificial drainage channels present in the lower part of the catchment to drain the valley peat and allow conifers to grow.

The storm of 5–6 August was uniform over both catchments and so, if the effects of vegetation are unimportant, the Severn hydrograph should show a quicker response because of the factors mentioned above. In fact, the Severn catchment proved to be slightly slower in response than the Wye; the peak flow was substantially lower and the recession limb less steep (Fig. 2.2).

These results are corroborated by an experiment in the USA. In the Fernow Forest, West Virginia, two very small catchments with the same forest cover were monitored for a number of years (Fig. 2.3). The hydrographs for this period are exemplified by 9–11 July 1955 when it can be seen that responses to rainfall are very similar, except that

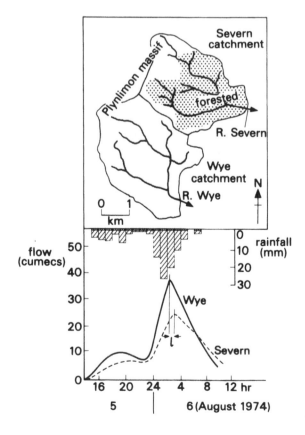

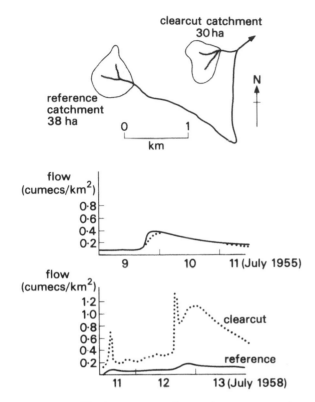

Figure 2.3 The importance of a continuous vegetation cover in runoff response.

Figure 2.4 A lowland area is much more liable to extensive flooding.

the peak flow of the control catchment was three hours ahead of the other catchment. After this monitoring period one of the catchments was clearcut (trees felled and removed) and natural regrowth of vegetation was allowed. A further monitoring period in 1958, when a partial grass and shrub cover had appeared over the clearcut catchment, revealed dramatic changes. For a rainstorm of 11–13 July 1958 (smaller than that in 1955) the control catchment (still with its forest cover intact) showed a response smaller than that for the storm of 1955. By contrast, the clearcut catchment responded very rapidly and appeared markedly more flashy (liable to rapid fluctuation) than before. In fact the clearcut catchment can be seen to have peaked some hours before the control, indicating a five hour reduction of the response time. Total yield was also found to be six times greater than the control catchment during the storm runoff period.

Lowland catchments

The runoff response of upland catchments is of particular concern because they can easily cause floods from high rainfall, but although lowland catchments usually have less rainfall of a lower intensity, there is still a real concern for flooding especially as they are generally more densely cultivated and populated. In lowland areas trees are less common, rivers have low gradients and erosional energy is reduced so valley slopes are less steep than in uplands (Fig. 2.4). In consequence soils are deep, making it more likely that infiltration capacity will not be exceeded. But because

valley floors are wider, the saturated zone adjacent to rivers is also broader and, as has been established, it is the zone nearest the river that is of particular significance in controlling overland flow. Thus the nature of lowland catchment response is a balance of the above factors and is best examined further by reference to an example.

The River Ter catchment is larger than those of Plynlimon, but over its 77 km² the range of altitude is only 80 m (Wye: 500 m range in 10 km²). Situated in the warmer regions of Essex, most of this catchment comprises cultivated deep, brown earth soils. Rainfall is typically of low intensity and Figure 2.5 shows that such a storm with 7·5 mm rainfall over a six-hour period took fourteen hours to give peak flow. Note however that a larger storm event of three centimetres rainfall in twenty-four hours produced a sharper hydrograph peak. Note too that the rising limb of the hydrograph becomes steeper. In this case the low intensity rain that fell early in the storm period must have filled the moisture deficit of the catchment, so the higher intensity rainfall that came towards the end was more likely to run off quickly, especially from the zone adjacent to the river. These are major changes in response and probably represent a greater contrast than for an upland area under similar conditions because the contributing area can expand more easily with low relief. After the storm runoff has

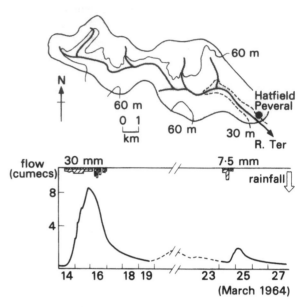

Figure 2.5 The response of a lowland catchment to medium and small storms; River Ter, Essex.

Figure 2.6 Contrasts in the response of the River Mole and River Kennet to storms in November 1974.

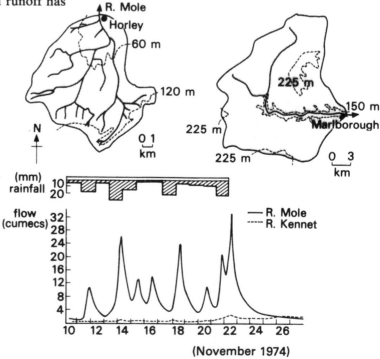

been taken away the river again falls to a low value commensurate with slow throughflow under a gentle gradient and, in this case, clay soils.

Finally let us examine the effect of a severe storm recorded on the River Mole catchment upstream of Horley, Surrey, between 11 and 22 November 1974 (Fig. 2.6). The River Mole has a similar amplitude of relief to the Ter (80 m over 89 km²) with impermeable bedrock. In this case a series of storms completely filled all moisture deficits so that the final 64 mm of rain culminated in a peak flow of 35 cumecs. For comparative purposes note that the hydrograph section on 11/12 November is similar to that of the River Ter for 15/16 March 1964. The rising limbs of all the stormflow hydrographs are strikingly steep and in this case are probably due to overland flow on the saturated zone beside the river channel. But equally striking is the rapid decay after each storm indicating that the base flow is characteristically very small. Indeed, at the end of the storm sequence the peak of 35 cumecs is reduced to a base flow of only 3 cumecs in only two days.

Conclusion: upland and lowland catchments compared Upland and lowland catchments can be contrasted in two ways. First, because of their effect on inducing precipitation, uplands tend to receive rainfall more frequently and with a longer duration than do lowlands. Because of the shorter time interval between storms, there is less time for soils to drain and so soil moisture contents remains high. Uplands therefore tend to have a lower soil moisture deficit to replenish before significant throughflow occurs. Secondly, steeper slopes in upland catchments promote faster throughflow which combines with the factors mentioned above to produce a smaller time lag between start of rainfall and start of runoff, as well as between peak rainfall and peak runoff, than is the case for lowland areas.

Saturated zones tend to reach their maximum development more quickly in upland than in lowland catchments partly because of the smaller zone involved and partly because of the smaller initial soil moisture deficit. Upland catchments tend therefore to experience overland flow more frequently and have more pronounced storm runoff peaks than do lowland catchments.

After the passage of a storm, runoff from uplands again declines more rapidly than from low-land because the steeper slopes encourage more rapid throughflow. Very low flows in upland environments are, however, mostly countered by the frequency of rainfall.

The physical effects of storm runoff are also very different on those occasions when the river channel cannot contain all of the water. Flooding in uplands is confined to a relatively small zone adjacent to the channel because of the steep slopes and absence of a floodplain (except in heavily glaciated valleys). As a result flood peaks are but little attenuated (p. 35) and they are transmitted almost intact towards the catchment outflow point. By contrast, the more gentle relief of lowlands, often coupled with a wide flood plain, results in considerable spillage, storage and attenuation of flow, so creating a less pronounced peak spread over a longer time.

The effect of permeable bedrock

Although rocks may be classified as permeable for hydrological purposes, there remains a considerable contrast between, for example, rolling chalk landscapes (Fig. 2.7) and the scar landscape of massive limestone (Fig. 2.8). Permeable sandstone strata add yet another possible landscape variation.

The nature of the response of any stream depends on: (a) how quickly storage can be filled up thus allowing an increase inflow; and (b) how much usable storage there is to sustain flow once rainfall has ceased. Before a stream can respond to rainfall in a catchment with permeable bedrock, a much larger storage capacity has to be filled than is the case with impermeable bedrock. As a result streams in such areas respond very slowly and exhibit none of the 'flashy' characteristics of the streams studied in earlier sections. The large storage capacity also results in more water being available for base flow, thereby giving very gentle decay limbs to hydrographs. This sort of response is shown by the hydrograph of the River Kennet above Marlborough, Wiltshire (Fig. 2.6), a catchment of 142 km² formed on the dipslope of the chalk. As with the River Mole, two major rainfall events are seen to be complemented by two rises in river flow, but in this case they comprise: (a) a small peak of limited duration immediately after each storm and probably due to expansion of the saturated zone in the alluvium either side of the river; (b) a progressive rise over and beyond the period of rainfall, very closely matched by the

Figure 2.7 Part of a chalk dry valley showing absence of surface drainage and 'rolling' topography.

Figure 2.8 The rugged landscape of Carboniferous limestone country, Malham, Yorks.

fluctuation in water level in nearby wells and thus due to ground-water contribution. Water falling over the majority of the catchment must have percolated deep into the fissured chalk where it supplemented the ground-water reserve already present. This caused a rise in ground-water level across the catchment such that after a period of time the water table intersected the ground surface over a progressively greater area, thereby producing channel extension into the dry valleys and augmenting the flow in the Kennet. In addition the rise in water table height near the catchment boundaries increased the hydraulic pressure on existing springs in the valley bottom, again augmenting the streamflow. This takes a long time because of the slow rate of transmission of water through the aquifer, and explains why the peak Kennet flow was eight days after the last peak due to stormflow. After this there was no further addition to ground water from percolation and the water table fell so that the Kennet flow began to decline slowly. With such a large storage component the possibility of flooding following storms is slight and this provides a marked contrast with the River Mole.

Chalk and sandstone are similar because water has to flow through many small fissures. However, inspection of a hydrograph for a massive limestone catchment (Fig. 2.9) shows considerable differences. For example, Carboniferous limestone of the Mendips and Pennines possesses large fissures which allow a response at least as flashy as a clay catchment. This is why pot-holers sometimes become trapped in the underground systems. Here one finds large impermeable blocks separated by fissures which have been much widened by solution and abrasion and are able to transmit water at a speed similar to open channels. Storage capacity is much less than in chalk. For example, the hydrograph for Bradwell Brook, Derbyshire, clearly shows how the brook responds strongly to rainstorms and, although all the water is 'ground water', the lack of storage results in considerable changes of 'water table' level, often of the order of 30 m. This occurs particularly during large storms when input from swallow holes exceeds the transmission capacity of fissures.

The amount of sediment produced by weathering of limestone is small – in such rocks the insoluble content is often less than five per cent. Because of this there is little to abrade the rock and erosion is mostly by solution, as is shown by the very high calcium, magnesium and bicarbonate ion levels (in this Derbyshire area about 300–450 mg/l). However, compared with other erosion systems, this is a small amount, which helps to explain why limestone rocks usually form elevated relief. Lastly, it must be remembered that, as with chalk areas, the ground-water catchment of limestone does not necessarily correspond to the surface topographic catchment, as it depends on rock structure and so comparisons based on surface catchment size must take this into account.

Conclusion: permeable and impermeable bedrock catchments compared It is clear that chalk catchments are dominated by a ground-water component, so that response to storm rainfall is slow. Similarly the decay limb is gentle and of considerable duration as water is released from ground-water storage long after the passage of the storm. Catchments with impermeable bedrock are, by contrast, quick to respond to rainfall as they have a low storage potential, giving large storm peaks, often causing floods and decaying rapidly to leave a small base flow generated from soil drainage only. Such contrasts are readily seen in Figure 2.10 where the hydrographs for the Kennet and Mole catchments are superimposed over two water years.

Catchments of snow and ice

Snow Whereas rain is the dominant form of precipitation in maritime temperate and tropical climates, polar and many temperate continental areas receive most of their precipitation in the form of winter snow. In extreme cases this snow may accumulate to form ice, but more commonly (as in the Canadian Prairies and Russian Steppes) snow falls each winter and melts in the following

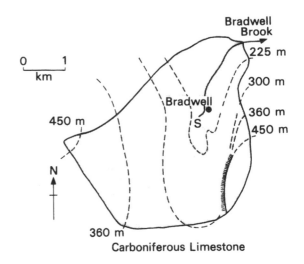

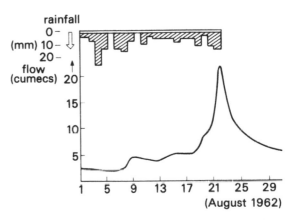

Figure 2.9 The Bradwell Brook near Matlock, Derbyshire. The hydrograph shows storage characteristics at first, but later responds rapidly to excess rainfall.

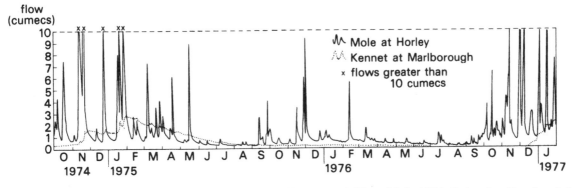

Figure 2.10 Long-term hydrographs for the River Kennet and River Mole 1974–7 showing flood and drought conditions. (Flows are averages per day.)

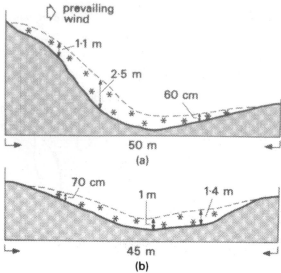

Figure 2.11 Snowcover changes with varying topography; Boggy Creek, Regina, Canada.

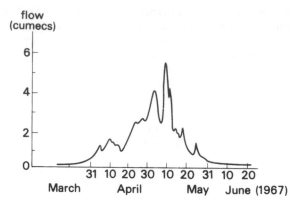

Figure 2.12 A hydrograph resulting from snowmelt, North Nashwaaksis stream basin, Canada.

spring. Because such regions are continental, their total water income is small and it is worth remembering that one metre of snow is equivalent to only eight centimetres of rain.

Snow is, of course, liable to considerable drifting so that it will accumulate anywhere sheltered from the wind. This may be a gully, in the lee of buildings or where there is increased surface roughness, as with stubble left standing in a field. This is illustrated in Figure 2.11 where a shallow valley cut by Boggy Creek dissects part of an open plainland near Regina, Canada. The greatest accumulation is, naturally enough, at the foot of steep sheltered slopes, the snow thickness declining rapidly towards regions of greater exposure (Fig. 2.11(a)). As the valley form changes and offers less protection, the snow thickness is again greatest in the shelter of the valley, but there is considerably less depth throughout (Fig. 2.11(b)). This sort of contrast is also to be found in forest belts, whether they be merely shelter belts or extensive tracts. In the forest zone of North America and the USSR, changes in the density of tree stands can materially alter the amount of accumulation and thus the potential for spring runoff. Here it was found that if the open space within the forest was too large, exposure to the wind reduced accumulation as on the open plain. In closed forests this was also the case to some extent as wind could blow across the upper surface of the forest taking snow from the upper branches. Accumulations were greatest in open forest where snow could fall between the trees rather than remain on the upper branches. In this respect, a stand of deciduous forest allows the greatest accumulation because there are no leaves to retain snow in the winter period (Table 2.1).

Table 2.1 The effect of vegetation forms on snow accumulation.

Area type	Fifteen year average of water equivalent as a percentage of birch forest accumulation
birch	100
pine	82
fir	63
small openings in forest	103
fields	68

Note the measurements were taken within a small area and precipitation in each area can be regarded as the same.

In the Canadian Prairies over 80% of the total streamflow often comes from snowmelt. The hydrograph characteristics of each river are thus determined in large measure by the depth of accumulation of snow and its distribution. When snow first melts, it is because of solar radiation; evaporation being unimportant as air temperatures are low. Snow melts and gradually replenishes the soil moisture deficit. Only after this does throughflow allow river flow to increase. Thus the hydrograph for the North Nashwaaksis Stream (Fig. 2.12) shows a slow rise from March to

April. After this, the effects of warmer air as well as radiation give a faster rate of melting and the peaks in April and May represent increased runoff due to the passage of warm air masses and occasional rainfall events. By mid-April the snow has largely melted and river flow has therefore declined by June. Note that the possibility of a flood – and consequent soil erosion – in these areas is dependent mostly on air temperature fluctuating rather than on rainfall.

In the forest zone to the north of the Prairies more snow accumulates for reasons mentioned above. The shade provided by the coniferous trees slows down melting from radiation, whilst a more northerly latitude causes the peak flow of rivers to come later in the year. As in other parts of the world, forested catchments are less liable to flooding than are the open plains.

Ice In many mountainous regions of the world, snow turns into ice to maintain the glaciers which linger in valleys as a last vestige of the Quaternary cold periods. Although these valley glaciers and ice caps are relatively small, they are sources of some of the most important rivers in the world and as such are worthy of consideration. In addition, much erosion was accomplished by fluvial action as ice sheets melted at the close of each cold period and thus a study of present ice-fed rivers may pro-vide an insight into this important aspect of geomorphology.

Most active glaciers are found in catchments with: (a) steep slopes; (b) thin, intermittent coarse debris cover on solid rock (moraine or outwash deposits) in areas without ice; and (c) very sparse vegetation (Fig. 2.13). These factors will obviously cause runoff to occur very rapidly after snow/ice-melt or rainfall begins, and the streams which flow from such catchments can be expected to be very flashy.

The normal sequence of runoff response begins in late spring following melting of snow. The first snow melted in May is largely absorbed by the underlying snow and refreezes. Soon this causes the snow to become more compact (then called **névé**) and less permeable. This in turn increases the probability that some further meltwater will have to run off across the surface. Meandering streams on the névé/ice surfaces of a glacier are very common (Fig. 2.14) and they connect together with englacial and subglacial streams to form the torrent that emerges from the glacier snout. In such a situation there is little storage and water velocities are high throughout. Rainfall is an additional contributing factor in two ways: first, in providing additional water and secondly by helping to melt the snow and ice. As the summer proceeds, the lower reaches lose their snow and névé cover and

Figure 2.13 The Engabreen glacier, Northern Norway, 1976.

Figure 2.14 Surface melting is quickly transmitted away by supraglacial streams.

Figure 2:15 Factors affecting the outflow hydrograph of the Engabreen catchment.

the exposed ice begins to melt. By early autumn, temperatures fall again to below freezing, ablation ceases and the precipitation is once more mainly snow so that streamflow rapidly diminishes. Thus, as with catchments of snow, the runoff regime is markedly seasonal.

The Engabreen glacier, Svartisen, is one of a group of valley glaciers in northern Norway which display the characteristics described above (Fig. 2.15). Of the 39 km² catchment area, 36 km² are permanently covered with ice and snow. During the eight-month period of snowfall there is considerable drifting, such that some exposed slopes remain snowfree, while more sheltered places accumulate up to 30 m. In general, however, snow depth increases with altitude whereas in summer ablation shows an inverse relationship, with the lowest areas melting first. At the end of the summer there may be some snow remaining on the uppermost part of the glacier, but because this is more than balanced by melting near the snout, this glacier suffers a net loss of ice over the year.

Runoff response of glacier-dominated catchments results from variations in air temperature, direct insolation and rainfall. The most important of these factors are rainfall and air temperatures because ice and snow reflect most solar insolation except towards the end of the summer when they may become covered by wind-blown dust. Rainfall will be seen to be the factor most directly responsible for producing runoff peaks. For example, the hydrograph peak of $5 \cdot 2 \times 10^6$ cumecs on 19 August can be closely correlated with 125 mm of rainfall which occurred on the same day. Over this period it will be noted that air temperature was fairly

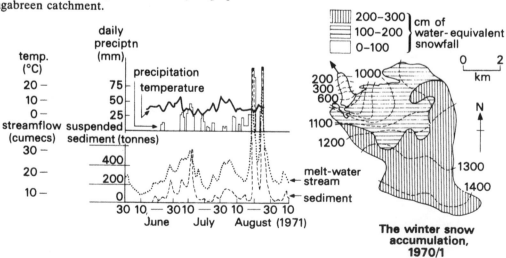

The winter snow accumulation, 1970/1

Figure 2.16 Wadis are characterised by considerable alluvial accumulations, resulting in flat floors.

constant with only a 2°C fluctuation. Although the effects of air temperature fluctuations are less dramatic, they are still clearly seen over the period 15 June–5 July. During this time the only rain to fall was on 22/23 June (25 mm), but the meltwater stream hydrograph shows a consistent rise throughout in parallel with the air temperature variation. The same is clearly true of the period 30 July to 8 August.

Conclusion Hydrological regimes of catchments dominated by ice and snow are distinctive primarily due to: (a) the large amount of water equivalent stored during the winter and made rapidly available in early summer; (b) the temperature dependence of the regime such that high flows can take place without precipitation; (c) the lack of storage elements within the catchment to retard response; and (d) the possibility that runoff can exceed rainfall due to glacier wasting.

Finally it will be seen that such large flows are able to move considerable quantities of sediment (Fig. 2.15) because the glacial streams are transporting agents for moraine produced by glacial erosion and made available at the snout.

Catchments of arid and semi-arid regions

Arid regions The main storage element in this type of landscape is found in the wadi (valley) bottoms where the flat floors are often mantled by a thick layer of coarse alluvium (Fig. 2.16). Debris on the hamadas (rock desert plains) has very little storage capacity and there is little fill in the bottoms of steep tributary valleys that make up the headward stretches of wadis (Fig. 2.17)

Figure 2.17 Areas above wadis are often steeply sloping and have little alluvial accumulation.

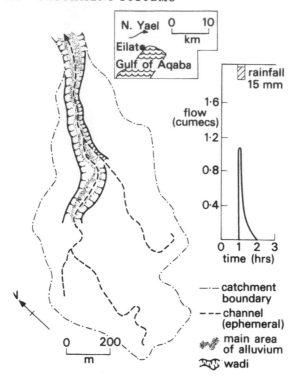

Figure 2.18   The result of a typical storm in the Nahel Yael catchment, Negev desert, Israel.

Surface-water movement in such areas can thus be divided into: (a) sheet flow over gentle hamadas which have very low storage potential; (b) cascading flow down the wadi sides; and (c) torrential flow or mudflow in the wadi bottoms. In all areas the moisture content of materials before a storm will be very low.

It is widely held that many of the major features of a rock desert are a result of action by water, although at the present time water is only available irregularly in very small amounts. This is the main reason for the suggestion that the landscape is essentially a fossil one and that it evolved under palaeoclimatic regimes wetter than that presently prevailing. The characteristic desert rainfall event is in the form of a thunderstorm lasting less than an hour and being separated from the next by some months. For example, while the Negev desert in Israel has an average annual rainfall of about 30 mm, within the period 1965–70 for example, no rain fell for two consecutive years.

The Nahel Yael catchment (nahel = wadi) (Fig. 2.18) is in the southern Negev desert. It is an area of 0·6 km² of more or less bare rock with a very

rugged appearance and many slopes in excess of 30°. The small tributary channels that begin in the upper part of the catchment lead into a major wadi, which eventually leads down to a closed depression (beyond the catchment boundary) a few kilometres north of the Gulf of Aqaba. Like many rock deserts it has only a thin veneer of material on solid rock and this is easily moved by flowing water. In the period 1965–70 there were only six storm events, each producing between 7 mm and 26 mm of rainfall. Although there is considerable variety in the character of storms in this area, a typical storm might consist of fifteen minutes of effective rainfall with a mean intensity of 0·5 mm per minute. In this catchment this value is close to the infiltration capacity of the wadi alluvial fill (which varies between 0·2 and 0·5 mm/min); thus storms with a mean intensity of less than about 0·5 mm/min produce runoff only on the upper parts of the catchment. When rainfall does exceed infiltration capacity over the whole catchment, flow is still generated in the headwaters first, whence it proceeds down the channel network as a true flash flood. It is even possible to see a wall of water moving down the channel, and this is responsible for the vertical segment of the hydrograph's rising limb. As the flood extends over the alluvium of the wadi floor, so it infiltrates and becomes much more attenuated, giving a total flow duration of less than a day and usually between one and four hours. The large infiltration capacity of the wadi alluvium results in flows out of this small catchment being fewer in number than the rainfall events.

As would be expected, one of the most dramatic type of flood occurs following intense convective thunderstorms whose duration and /or direction of movement are such as to maximise the runoff in the catchment (p. 15). Such storms are rare, but one occurred in Eldorado canyon, a small barren and rugged catchment of 59 km² about 80 kilometres south-east of Las Vegas, Nevada, in 1974. At about 1 p.m. a thunderstorm broke over the upper part of the catchment. Rainfall was heavy, exceeding 75 mm per hour over the whole storm period and exceeding 150 mm per hour for half an hour. The storm then swept down the catchment at about 12 km per hour coinciding with the movement of the storm runoff. This caused intense rainfall and runoff to be superimposed, thereby compounding the peak intensity of surface runoff. The tendency of the front of the flood wave to pick up debris from the dry stream channel ahead of it made it move slower than the water behind. The initial surge of water that swept down the canyon was described as looking like freshly mixed concrete with water piling up behind it all the time. As a result the flood wave arrived as a wall of water and debris at the lower end of the catchment. The lower reaches are used as a recreation resort and there were several buildings in the canyon. One of the tourists in the resort described the arrival of the flood wave:

When I got around the small nose which was behind the block-ice machine and started walking towards the coffee shop, I looked up for a second. I became disoriented because I thought the mountain had moved. Then I realised what we were seeing was a wall of water about 20–25 feet high stacked with cars, trailers, etc., smash into the coffee shop, post office, and they exploded like there was dynamite inside.

In this case the major damage and the loss of life were caused by the leading edge of the flood wave which was moving at about 4 km per hour. The whole storm was past by 2.30 p.m., a total duration of 1½ hours, but during this time nine people lost their lives, buildings were destroyed and 70 000 m³ of material was removed from the catchment, mainly in the first few minutes, lowering the landscape by an average of 0·6 mm. The average sediment content of the torrent was calculated at 35 000 mg/l.

Semi-arid environments   The transition from arid to semi-arid is characterised by: (a) an increase in the frequency of convective storm events; (b) river flow that can be maintained for periods of weeks or more; and (c) the survival of drought-resistant vegetation such as mesquite, cottonwood and prairie grass. Despite this the effective precipitation is low, vegetation discontinuous and streamflow characteristically very irregular (Fig. 2.19).

A large part of south-west USA experiences semi-arid conditions and the heavy storms which are a feature of such climatic regimes. *The Chronicle News*, Trinidad, Colorado, reported the results of one storm sequence as it affected a small tributary of the Purgatoire River:

Thursday (17 June 1965) the water was running about normal, or maybe a bit higher. I had started to read *The Chronicle News*. ... I got curious about the Creek (Raton Creek at Starkville, Colo.), it was raining you know. I had pulled the blinds down and then I got up to look out the front door, and there it was coming up as fast as everything. By then it was already around the house. It started coming in under the door and I knew it was going to be bad, but by then it was too late to get out.

I tried to call the sheriff and I talked to someone who couldn't understand who I was, but I told him ... that I needed some help to get out. ... When I placed the phone back on the stand, I

Figure 2.19 Semi-arid areas differ from arid areas in the amount of vegetation present and the duration of streamflow. Note the straight slopes and sharp crests.

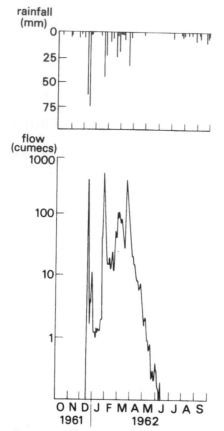

Figure 2.20 Flow in Sycamore Creek is often continuous for part of the year.

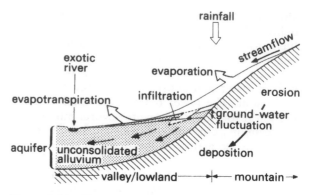

Figure 2.21 The water balance of a semi-arid region. The maximum streamflow occurs at the lowest part of the mountain section.

was in water almost up to my waist. ... Right then with the water pouring in and the noise and everything, I really thought I was a goner.

I put my dog on the bed and I got up there too. I held tight on the head of the bed and it started jumping around, floating here and there. The water just kept coming in and filling things up. I was afraid it would get so high I would drown.

Just about then, when I thought the water was going to fill the whole house, a part of the wall right in the bedroom broke, and then water was running out as fast as it ran in. I could hear windows breaking and doors were crashing and furniture was swimming around. When the wall broke out I thought maybe I can still save myself after all. ... I began to get awful cold and was shivering wet. It smelled awful, that mud and dirt in the water.

It is clear from this description that there are few storage elements in the landscape to slow down the storm runoff. The major storage element, as with wadis, is the valley alluvium that accumulates because streams rarely last long enough to carry it away. However, when a sequence of storms affects a catchment, this storage is rapidly filled. The number of times the storage was filled for Sycamore Creek near Phoenix, Arizona is shown in Figure 2.20, and the effect of valley alluvium on the hydrograph in Figure 2.21.

Conclusions Arid and semi-arid catchments are characterised by intermittent runoff generated by infrequent storm events. Semi-arid regions are distinguished from arid regions primarily by the frequency of rainfall and the amount of vegetation that can survive. For an individual storm event hydrographs have very steep rising limbs and steep decay limbs. Because of the limited nature of runoff, material in transport from the upper slopes is deposited when the gradient decreases resulting in more depth of alluvium and somewhat greater storage for runoff. In semi-arid areas such storage is filled more often so that, if rainfall continues, flooding occurs in the lower valley reaches.

Humid tropical catchments

Probably the most distinctive feature of humid tropical areas is the seasonality of rainfall. In most places having such a climate one or two wet seasons are separated by dry periods in which virtually no rain falls. This means that river flows show an extremely large seasonal fluctuation which often overrides the effects of individual storms. In addition the high temperatures (annual average above 25°C) increase the rate of chemical weathering such that it may be one hundred times faster than in humid temperate regions. Because it is usual for rainfall to exceed evapotranspiration in the wet seasons, the combination of high rainfall and temperature promotes the disintegration of rock into clay minerals, which may reach depths of tens of metres (see p. 49). The deep clay soils result in slow percolation rates and, although much water does percolate into the soil, there is a high probability of the infiltration capacity being exceeded so that overland flow may occur during storms, although mainly in the root mat of the forest cover and over swamp area. When the seasonal rains are over, the upper soil quickly dries out and potential evapotranspiration may be of the order of ten times the total runoff, leaving only deep throughflow and ground-water flow to maintain water in rivers. As a result the flow quickly reduces to a low fairly constant value throughout the dry season, as is reflected in the hydrograph for the River Sonjo, which lies nearly 8° south of the equator in the tropical forest belt of East Africa (Fig. 2.22). The 68 km² catchment area is typical of many in tropical Africa in that it comprises gentle slopes of ancient peneplained areas, occasionally dissected by rivers (Fig. 2.23). The majority of areas between valleys have very low hydraulic gradients and water retained in the deep soils at the end of

the rainy season is probably lost more by evapotranspiration than by lateral transfer to rivers. Rivers are in fact mostly sustained by throughflow from slopes fairly close to the channels. As in other areas, shallow hillside concavities are the main sources of overland flow. These areas are probably too wet to support trees and so are seen as clearings in the forest. Such source regions quickly replenish their moisture deficit at the start of the wet season, so that by mid-March much of the rainfall runs straight off and it would seem that this is the main cause of the hydrograph spikes.

In areas such as this, hydrological parameters are difficult to measure and rainfall values are not readily available. Nevertheless rainfall appears to come in a series of well-defined storms as represented by simple peaks in the wet season hydrograph. The similarity of this to the more extreme rainfall events on the Mole (Fig. 2.10) is striking. Floods are thus common features of tropical rivers, but the deep and intense weathering robs rivers of the coarse bedload they would use in temperate regions to erode their channels so that landscape incision is less effective than in temperate regions.

The influence of urban areas

Urban development creates particularly drastic changes in catchment response because it not only

Figure 2.22 Flow conditions for the River Sonjo 1964–5.

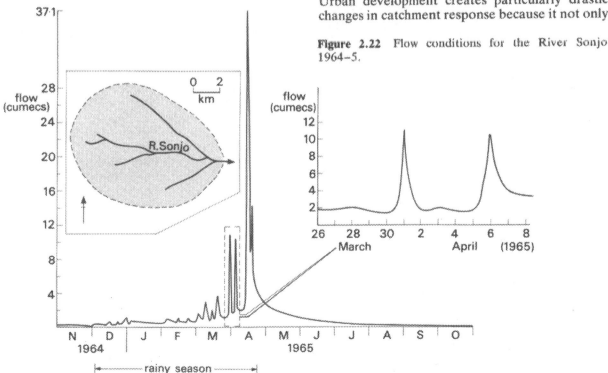

Figure 2.23 The humid tropics are characterised by luxuriant vegetation growth and deep soils. However in many areas this has been cleared for cultivation and secondary regrowth dominates.

Figure 2.24 Urban functions cause changes to stream channel as well as sources of water inflow; River Tame, near Wednesbury.

changes land use and the nature of cover but also reorganises the drainage network and sometimes hillside slopes. With 80% of the population of many developed countries now living in urban areas, the responses and effects of urban catchments demand particular attention. Nevertheless storm drainage systems in cities today are still largely designed on empirically derived formulae – a hazardous procedure when one considers the enormous expenditure involved.

The urbanisation process is familiar: initially rural lands are colonised by villages whose growth at first has little impact on the environment, but as they develop into towns and cities two things happen. First, the centre becomes increasingly more densely occupied – often to 100% cover of the land. Secondly, with increasing pressure on central land some people seek more space by moving to the outskirts of the urban area and thereby establish a suburban fringe. In the suburbs, housing density is less, with larger gardens, but locally there can also be complete cover as neighbourhood shopping centres and car parks are developed. As a result, in many areas there is a gradually rising percentage of land affected by urban functions (Fig. 2.24).

Most of the hydrological consequences of urbanisation are apparent. Impermeable surfaces are drained by artificial channels (storm drains) which have smoother bed and banks than a natural stream channel. Similarly the lower friction of drains allows water to move faster and promotes rapid runoff. In completely urbanised areas there may be no storage component at all because water is prevented from entering the soil. Thus a stream affected by urbanisation would be expected to have a shorter lag between rainfall and runoff, a higher storm peak and a much reduced base flow component compared with the response of the same stream under natural conditions. Moreover, with water going into the drains and not being available for evapotranspiration or replenishment of soil moisture storage, a higher total water yield from rainfall would also be expected.

Beyond the city centre the suburbs cause less pronounced changes because there are more open spaces whose hydrologic response to rainfall is little affected by urbanisation. Nevertheless between 20 and 50% of rainfall will still fall on impermeable surfaces and much that falls on vegetated ground will also find its way to tile and storm drains, so again water yield will be increased and the hydrograph will appear more peaked than for a natural catchment.

Some of the effects described above can readily be seen in the hydrographs for Canon's Brook, Harlow New Town, Essex. Harlow was developed as one of London's overspill towns and its population grew very rapidly between 1953 and 1971 from 6000 to 78 000. The major part of the town is drained by a small stream, the Canon's Brook (Fig. 2.25(a)). The 21·4 km² of this catchment was totally rural in 1953, but fifteen years later it had changed to 17% urban. Over the period 1950–4, before urbanisation had got underway, a series of storms were recorded and the unit hydrograph calculated. Time from start of rise to peak was about 5 hours, in marked contrast to the unit hydrograph for storms between 1966 and 1968

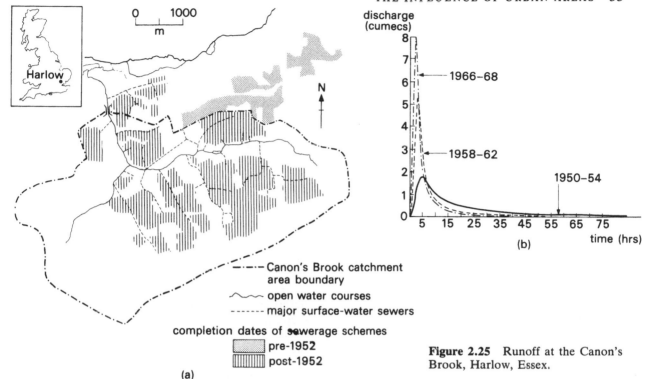

Canon's Brook catchment area boundary

open water courses

major surface-water sewers

completion dates of sewerage schemes

pre-1952

post-1952

(a)

Figure 2.25 Runoff at the Canon's Brook, Harlow, Essex.

whose time to peak was only 2½ hours (Fig. 2.25(b)). The same sort of pattern has been recorded for Crawley New Town, where the urban area for a very small stream, the Crawter's Brook (2·2 km²) increased from 5% to 26% in the years 1949 to 1969 and the time of rise was reduced from 2½ to 1½ hours.

Note that there seems to be a larger change in catchment response than might be expected from such relatively low levels of urbanisation. This is probably due to the stream realignment and concreting of channel banks and bed which occurs at an early stage of development. Water can thus move faster along the same watercourse because of less frictional resistance. In addition major storm drainage networks were installed in each case and the drains fed directly into the brooks. This is confirmed by looking at a catchment of 4·7 km² adjacent to Crawter's Brook whose urban area was only increased from 18% to 27% following the installation of a new storm water drainage system.

Crawter's Brook is a small tributary within the River Mole catchment, but the effect of urbanisation on the Mole catchment is usually not very marked because Crawley is a small part of the total catchment. However the effects of Crawley can be seen very clearly at times of very low water in the

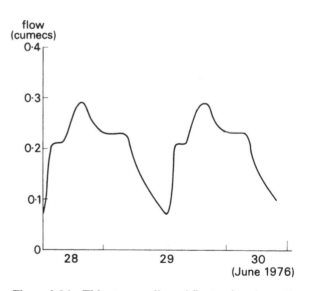

Figure 2.26 This strange diurnal fluctuation shows the effect of discharges from the effluent disposal plant (serving Crawley) on the dry weather flow of the River Mole at Horley in 1976.

Mole. Natural low flow from this clay catchment is small and so the daily discharge cycle of the town effluent disposal plant shows up as a distinct fluctuation (Fig. 2.26). The concepts involved in the discharge of effluent into rivers as the result of urbanisation are further discussed on p. 71.

The same effects of urban development on storm response can be seen just as clearly on larger catchments. The Brays Bayou catchment in the city of Houston, Texas (population 1·3 million) has an area of 350 km² ranging from a heavily urbanised core (impermeable area approximately 100%) to suburbs with impermeable areas of less than 50%. As at Harlow and Crawley this catchment became progressively urbanised by the south-western spread of Houston between 1939 and 1960. As a result the hydrograph changes are similar but of course the time scale and flow peaks are different (Table 2.2). Again the effect, not just of impermeable areas, but of the efficient storm drain networks has increased the peak flow threefold and reduced the time to peak flow by a factor of four.

Table 2.2 Hydrograph parameters for the Bray's Bayou catchment, Houston, Texas.

Date of storm	Peak discharges (cumecs)	Time to peak (hrs)
July 1939	58	12
May 1953	71	6
April 1959	147	4
June 1960	147	3
predicted ultimate	194	2½

In many cases it has been necessary to undertake major river works to handle greatly increased peak flows. Dykes are the usual answer for large flows but an alternative, or supplementary solution where practicable, is to divert some of the storm water to large 'soakaway' pits (recharge basins). This is a scheme operated, for example, in Nassau County, Long Island, New York State, where over one hundred soakaways are used for an area of 25 km² of suburban development with water soaking down to the ground-water reserves of the local aquifer. On a quite different scale many suburban areas of Britain, without main storm drains, have a statutory requirement to build small soakaways of about one cubic metre beside each house into which the rainwater pipes feed. Clearly such water must be kept away from effluent disposal sewers: (a) because the treatment plants are not designed to handle large water volumes as would occur with peak flows; and (b) because of the risk of sewers flooding out into streets and causing risk of infection.

Probably the greatest all round effect of urbanisation can be seen in the semi-arid South-West of USA. Here high water flow is modified from the natural regime in the same way as described above. Low water flow is seen as a diurnal fluctuation because of effluent discharge, but in addition irrigation of lawns and flower beds in the dry season has meant that many once ephemeral streams are now perennial.

Larger catchments

We have seen how climate, slope and lithology interact in determining runoff response. However, all but the smallest catchments contain significant variations in one or more of the factors so far described and the final river hydrograph is correspondingly complex.

As catchments increase in size so does the difficulty of isolating the factors that cause runoff response. For example, as a storm moves across a catchment so rainfall varies in space and time to an extent that it is usually no longer possible to relate rainfall at a particular site in a large catchment directly to the outflow hydrograph. The time taken for runoff from headwaters to reach the outflow point may be significantly different from the lag time of tributaries much closer to the outflow. The final hydrograph is thus a summation of the responses of each catchment section suitably lagged, and the larger the catchment the longer the lag, from an hour or two in very small catchments to a few days for rivers like the Severn, more than a week for the Rhein and even several months for the Niger.

Most large catchments have extensive flood plains. These are very important hydrologically because of their attenuating effect at times of flood. High flows generated in tributaries may produce a combined flow which exceeds the capacity of the main channel, causing overbank spillage and bringing the flood plain into use as a major storage element. Water moves much more slowly over a flood plain than in the river channel because of the retarding effects of vegetation and shallowness of water. For just a few centimetres rise of water level the flood plains can also store very large amounts of water. This returns to the river only slowly because the slope across the flood plain

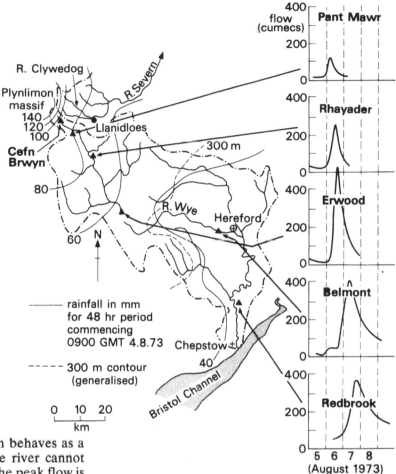

Figure 2.27 A prolonged storm in August 1973 was mostly confined to the mountainous area of Wales and reached a maximum intensity early on 6 August (see Fig. 2.2). The storm runoff can be seen to increase as far as Erwood. Below this no tributaries enter from mountainous areas and the flood wave is attenuated as it passes Belmont and Redbrook.

is so small. In effect the flood plain behaves as a natural reservoir for water that the river cannot deal with immediately. As a result the peak flow is reduced, while the length of time for which the land is flooded is increased (this is called **attenuation of flow**). Sometimes this effect can be clearly seen as was the case for the storm of 5–6 August 1973 which fell only on the headwaters of the River Wye in Wales (Fig. 2.27). The catchment lies on impermeable rocks throughout, so the flood generated in the headwaters was transmitted downstream as a distinct pulse. The delaying effect of flood plain and channel storage can clearly be seen, with the peak gradually being flattened downstream so that at Redbrook near Chepstow it was not received until two days after the storm. In a similar way snowmelt flooding in the Alps can be traced downstream to Rotterdam on the Rhein despite the many source regions drained by tributaries of this large river (Fig. 2.28). Note however that individual storm events such as those traced on the Wye would not be readily recognisable with this scale of catchment. In the case

of snowmelt there is a prolonged late spring and early summer flood which is attenuated in form and lagged by three days on its passage from the upper Rhein at Rheinfelden to Cologne at the start of the North German Plain. Snowmelt effects are particularly clear here because rainfall is low over the catchment at this time. At other times of the year the correspondence between Rheinfelden and Cologne is less clear because of varied contributions from rainfall in tributary areas.

The Purgatoire river system, Colorado, provides an example of how hydrographs become more complex as the number of contributing areas increases. In this case the headwaters were dominated by storms near Raton (Fig. 2.29) which produced a double peak on 16 June at Hoehne. Because there was only one source region this storm was transmitted down the river to Alfalfa and then Las Animas as a simple wave with only the hydrograph shape changing due to river

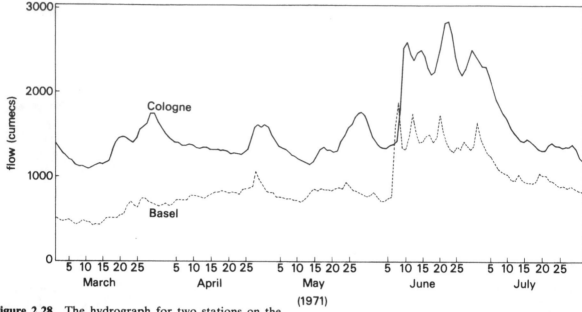

Figure 2.28 The hydrograph for two stations on the Rhein, Germany in 1971 showing the passage of the Alpine snowmelt peak.

attenuation (p. 35). However on 17 June storms were centred not only at Raton but over other tributaries in the south-east of the catchment as well. Consequently the simple peak at Hoehne has increased and become a double peak because of further contributions between Hoehne and Alfalfa. Another large contribution below this point again changes the hydrograph shape considerably so that the simple storm at Raton is now hard to find in the hydrograph of Las Animas.

Example problems

The purpose of this chapter has been to show how the principles in Chapter 1 are reflected in real situations. The examples chosen are of small areas where the number of variables in any one case is limited, but it is important not to conclude without reference to large river systems because these are the most familiar and important to us.

A: The River Thames catchment above Teddington, West London The Thames catchment, small by world standards at 9950 km², is still one of the largest in Britain

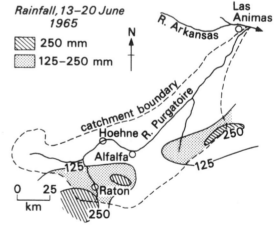

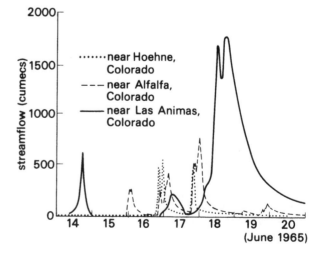

Figure 2.29 Hydrographs from selected stations on the River Purgatoire, Colorado and the storm rainfall pattern for the catchment; June 1965.

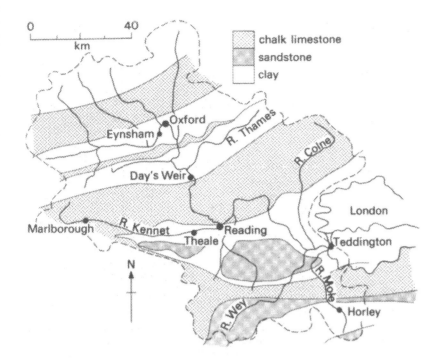

Figure 2.30 The Thames catchment above Teddington.

(Fig. 2.30). Precipitation is predominantly from frontal systems passing west to east, but the hydrographs at Teddington (Fig. 2.31) are more a reflection of contributions from subcatchments of varying lithologies than variations in rainfall pattern.

1. Prepare a list of the possible contributing sources reaching the river at Teddington based on the theory of Chapter 1 and Figure 2.30.
2. Compare the flow records for November and December 1974 at Eynsham, Day's Weir and Bray with that of Teddington (Fig. 2.31). Is there a clear progression of a flood wave down the river? If not, could this be due to: (a) time lag; (b) a particular rainfall pattern; or (c) the effect of variation in catchment lithology? Discuss each factor in turn.
3. (a) Refer to Figure 2.6. Does the time of peak of the Mole help to explain the peak at Teddington?
   (b) If it takes water about half a day to flow from Bray to Teddington, what is the greatest contribution (in cumecs) that can be expected from the middle and upper Thames at the time of the Teddington peak?
   (c) Calculate the difference which must come from contributions below Bray.
   (d) The Mole's peak contribution was about 102 cumecs at its confluence with the Thames. Try to identify other rivers that might make similar contributions using Table 2.3 to help calculate peak flows.
   (e) After adding these contributions together is there still more water needed to give the peak at Teddington? If so, where might it come from?

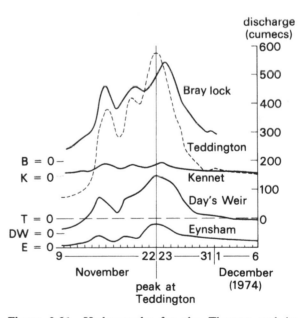

Figure 2.31 Hydrographs for the Thames and its tributaries, November–December 1974.

Table 2.3 Percentage contributions of the Thames and its tributaries at high and low flow.

Tributary	Percentage contribution at high flow	Percentage contribution at low flow	Distance from Teddington (km)
Windrush	4·1	5·6	173
Thames at Eynsham	19·7	14·6	160
Evenlode	5·9	3·5	157
Cherwell	12·6	3·5	150
Ock	2·8	2·5	136
Thame	8·7	2·4	123
Thames at Goring	53·3	29·1	106
Kennet	8·5	23·5	86
Loddon	7·3	13·0	79
Thames at Bray	73·8	71·8	46
Colne	6·7	7·0	27
Wey	6·5	10·7	19
Mole	6·3	5·0	8
Thames at Teddington	100·0	100·0	0

4. Table 2.3 shows the percentage contribution to high and low flow of various tributaries to the flow at Teddington. Plot a graph of distance above Teddington against contribution: (a) at high flow; and (b) at low flow for the stations listed. Using this graph identify areas with high and low ground-water contributions.

5. Examine the average monthly rainfall and flow of the Thames from Table 2.4. Plot the average monthly rainfall as a histogram and on it also plot the flow. Shade in the area between the two histograms. What is the chief factor responsible for the misfit between rainfall and runoff patterns and why does the difference vary over a year?

Table 2.4 Monthly average rainfall and runoff for the Thames catchment above Teddington, 1941–70.

Month	Rainfall (mm)	Runoff (mm)
J	65	36
F	48	33
M	47	31
A	47	22
M	59	17
J	52	12
J	60	9
A	71	8
S	63	10
O	66	14
N	76	23
D	69	31

B: The River Niger catchment above Onitsha, West Africa The Niger catchment (Fig. 2.32) area is one hundred times the size of the Thames, with a river length of some 4400 km. The effect of tributaries will be seen in so far as they drain regions of different climate; on this scale, lithological influences are of very limited significance. Near the deltas the many tributaries, which drain regions as contrasting as desert and rainforest, combine to give a flood wave which lasts for over a month. Individual storm characteristics are lost on this scale and the hydrographs of the lower Niger reflect seasonal variations in precipitation effectiveness rather than storm events. Nevertheless the individual tributaries (each larger than the Thames) do respond to individual storm events, as will be discussed below. Ultimately much of the industrial potential of West Africa will depend on the proper use of the resources that this river can provide: already there is a huge dam at Kainji impounding a lake 136 km long and there are more to follow. It is therefore especially important that the hydrological responses of the whole catchment are understood so that the most useful watershed management practices can be employed. The data used in this section refer to the period before the Kainji dam's construction. The influence of the dam on the Niger flows is discussed on p. 73.

1. Examine Figure 2.32 which shows the variety of climatic regimes experienced within the catchment. It is estimated that the total annual flow of the River Niger as it leaves the mountains of Sierra Leone is about 45 km³, while a tributary, the River Diaka, contributes about 40 km³. In both cases peak flow occurs in August. (a) Contrast this information with the hydrograph for Diré (Fig. 2.33(a)). (b) The hydrograph is notable for its smoothness. What must this be due to when the rainfall pattern shows great seasonal contrast? (c) The area between the mountains of Sierra Leone and Diré has an annual rainfall of about 250 mm and a potential evaporation of 2500 mm. The total annual flow of the Niger at Diré is about 40 km³. Does this tell us anything about the size of the 'lake' above Diré? (Refer to the flows of Question 1.)

2. Downstream of Diré the river flows in a well-defined wide alluvial channel into regions of greater rainfall. Nevertheless the hydrograph peak produced at Diré (known as the Black Flood) is identifiable all the way down to Onitsha. Why should this be so? (Examine the rainfall regime of the regions between Diré and Onitsha.)

3. Contrast the Niger hydrographs at Jebba and Lokoja. The peak flow at Lokoja occurs before that of Jebba and, more noticeably, the seasonal hydrograph rise begins much sooner. Identify the tributaries that cause these high flows and attempt to relate the observations to seasonal shifts of the equatorial low pressure belt (Fig. 1.2).

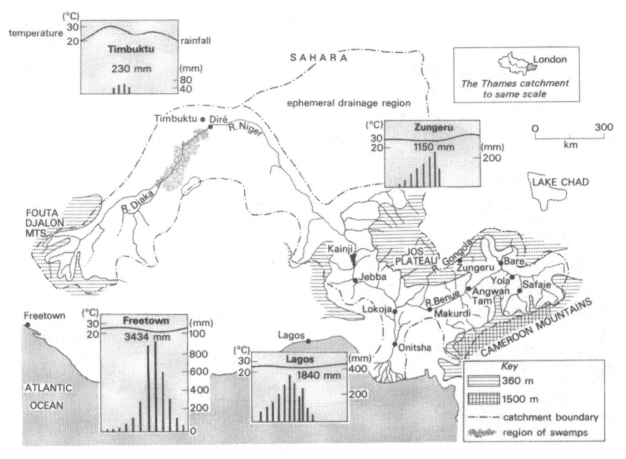

Figure 2.32 The Niger catchment, West Africa.

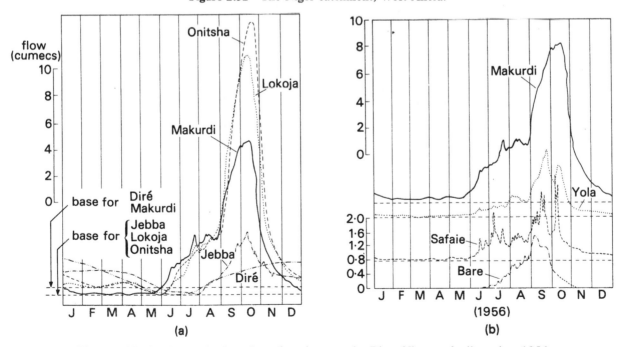

Figure 2.33 Hydrographs for selected stations on the River Niger and tributaries, 1956.

4. (a) The Benue is the most important of the River Niger's tributaries. It has headwaters draining the tropical rainforest of the Cameroon mountains as well as the Jos Plateau. How would you account for the difference between the hydrographs at Safaie and Bare?

(b) Below Yola the river gradient is only 100 m in 500 km, the river cannot carry all of its coarse load derived from the headwaters and so the Benue flows over considerable depths of alluvium. How is this reflected in the hydrograph at Makurdi?

(c) Why does the Benue contribute so little in the period January to March? (Look carefully at the climate graphs.)

(d) Despite a long period of low flow and a catchment only half as big as the Niger above Lokoja, the total annual flow for the Benue is about the same as the Niger at their confluence. What factors might account for this?

5. Why are there no flow-measuring stations below Onitsha?

C: The comparison of hydrographs   There is a great need for within-catchment and between-catchment comparison of hydrograph forms so that accurate assessments of catchment responses may be obtained. Throughout this chapter the raw data have been presented, but with sufficient information a unit response graph type of standardisation can be attempted. The data needed for the River Mole are given in Table 2.5.

1. On graph paper plot the three unit response graphs of Table 2.5 in the same form as Figure 1.18.
2. Draw a separate graph (similar to the inset on Fig. 1.18) for unit response graph duration against peak flow.
3. What are the differences between the unit response graph set for the Rosebarn catchment (Fig. 1.18) and the Mole catchment and to what may such differences be attributed?
4. Plot the duration/peak data for the Rosebarn catchment on to the graph of the Mole. (a) Why are these two lines displaced? (b) Why do they have different slopes? (c) Is there any significance to extending these lines beyond the regions containing the plotted points and how far could such extensions meaningfully be taken?
5. Derive unit response graphs for some of the catchments in Chapter 2 using the method of p. 15 and compare them with response graphs of similar duration from the River Mole.

D: A fieldwork problem   An experiment was set up to investigate whether there was a significant variation in throughflow response between a strip of steep hillslope vegetated with coarse grass and bracken and an adjacent strip planted with larch. The experiment was conducted

Table 2.5 Unit response graphs for the River Mole above Horley.

5 hr, January 1969

Time from start of rise (hr)	flow (cumecs)
0	0
5	5
10	20 (peak)
15	12
20	6
25	3
30	1
35	0

10 hr, January 1970

0	0
5	5·5
10	14
12	15 (peak)
15	12
20	8
25	5
30	3
35	1·5
40	0·5
50	0

27 hr, June 1971

0	0
5	1
10	4
15	8·5
17	9 (peak)
20	8·75
25	7
30	5·5
35	4
40	3
45	2
50	1
55	0·5
60	0

between 14 July and 21 August 1975, in the Twyi forest, Dyfed.

1. How important would evapotranspiration effects be on soil moisture content at this time of year? Would the soil tend to be wet or dry at the start of each storm event?

2. Throughflow was measured at the eight places shown below for an apparently uniform silty clay soil (an acid brown soil of p. 44). The cumulative results were:

Grass/bracken slope		Forested slope		Slope angle	Distance from stream
A1	3 ml	B1	259 ml	28°	450 m
A2	8 ml	B2	273 ml	28°	300 m
A3	51 ml	B3	871 ml	40°	150 m
A4	42 ml	B4	4×10^6 ml	16°	20 m

What are the main contrasts in flow pattern: (i) between A and B; (ii) within A and within B. From these data can the object of the experiment be supported?

3. The permeability of the soil at each site can be calculated from Darcy's law:

$$Q = K \times A \times \tan \text{ (slope angle)}$$

where Q = flow rate in ml/s, A = area of throughflow measurement (in each case about 150 cm²), K = saturated permeability in cm/s.

Calculate the average saturated permeability of B4 over the duration of the experiment.

A sandy soil would have a maximum saturated permeability of 0·02 cm/s. Compare this with the value you have just calculated for B4.

4. By reference to Figure 4.18, suggest what the throughflow pattern must be near to B4 to give it such a large flow.

N.B. In such situations sufficient flow cannot be attained within normal soil pores and we often find that gaps between soil structural units (Fig. 3.10) have been enlarged by flowing water to take the required volume. In these situations water is virtually flowing partly in a series of subsurface pipes (p. 17). This **pipeflow** (which is of restricted extent) may come to the surface as a spring or continue directly to the streambank.

PART TWO APPLICATIONS

The hydrological cycle plays a fundamental role in our natural environment. In the following chapters we consider some of the more direct influences of water movement on the development of soils and valley slopes to illustrate this role. We conclude with an examination of how man attempts to modify the hydrological cycle for his own needs. Each chapter presents a hydrological slant to topics often covered from more conventional standpoints and may be read in conjunction with the standard treatments available.

Chapter 3

SOIL DEVELOPMENT AND MANAGEMENT

Soils of temperate climates

Hillside soils In a freely draining soil as water moves under the influence of gravity (Fig. 3.1) it performs a variety of roles: (a) it reacts chemically with the mineral and organic constituents of the soil (**weathering**); (b) it transports soluble products from upper to lower soil horizons and from upper slope elements towards the valley bottom (**leaching**); and (c) it transports the smaller insoluble particles such as clay minerals from upper to lower soil horizons and from upper to lower parts of a hillside (**eluviation**) thereby altering the permeability of the soil (Fig. 3.2). The result of these processes is that soils near the top of slopes tend to be acid with a deficit of fine particles and those at the foot tend to be much less acid and have a higher proportion of fine particles.

Figure 3.3 shows a sequence of soils developed on a uniform parent material of soliflucted shales. Near the top of the slope, **podzol** soils have developed with an average acidity of pH 4·5, but this gives way downslope to **acid brown** soils with a pH of 5·5. The acid brown soil continues until near the slope base, where a change of gradient causes basal waterlogging of the profile and finally peat development. With parent material, climate and grass cover all uniform, the variations of soils must be entirely the result of throughflow and percolation. Here **translocation** (leaching plus eluviation) on the upper slope has made the soil so acid that iron and aluminium have been mobilised and transported down the profile to leave an ash-coloured Ea horizon. Throughflow must be continually taking soluble weathering products from this part of the slope down to the river, but *en route* such flow must pass through the acid brown soils. Here the same weathering takes place, but part of the loss is made good by input of cations from

upslope and so the soil never becomes as acid as the podzol, iron and aluminium are not mobilised and the profile remains a uniform brown colour.

Although soluble products of weathering may travel a long way down the slope before being precipitated or may even be washed away altogether, it is rare for the insoluble products, like clays, to move very far. The movement is, nonetheless, a result of throughflow patterns. Redeposition of clay particles in B horizons mainly takes place along the sides of the soil pores making them smaller and thereby reducing permeability. This can be clearly seen in a thin section of soil (Fig. 3.4).

Flood plain soils Soils near to streams differ from those of hillslopes in that they are permanently saturated at depth and often saturated right to the surface in times of flood. Such conditions cause lower soil horizons to become a grey colour and upper horizons to show a mottling of grey or blue with orange or yellow. Many flood plain soils have these typical characteristics and are called **ground-water gley** soils (Fig. 3.5).

The main process involved is called **gleying** and results from the inability of soil organisms, such as most bacteria, to find enough oxygen in the soil water or from the very limited supplies of soil air that exist under saturated conditions. On a flood plain, water movement may be very slow except immediately after storms because of the low hydraulic gradient, and oxygen is consumed more rapidly than it is supplied. Because of this, oxygen has to be acquired from other sources such as iron oxides. In air, iron oxides help to give a soil a brown or orange colour but, with the loss of some oxygen, the iron changes from the ferric to the ferrous form typified by a grey or blue colour.

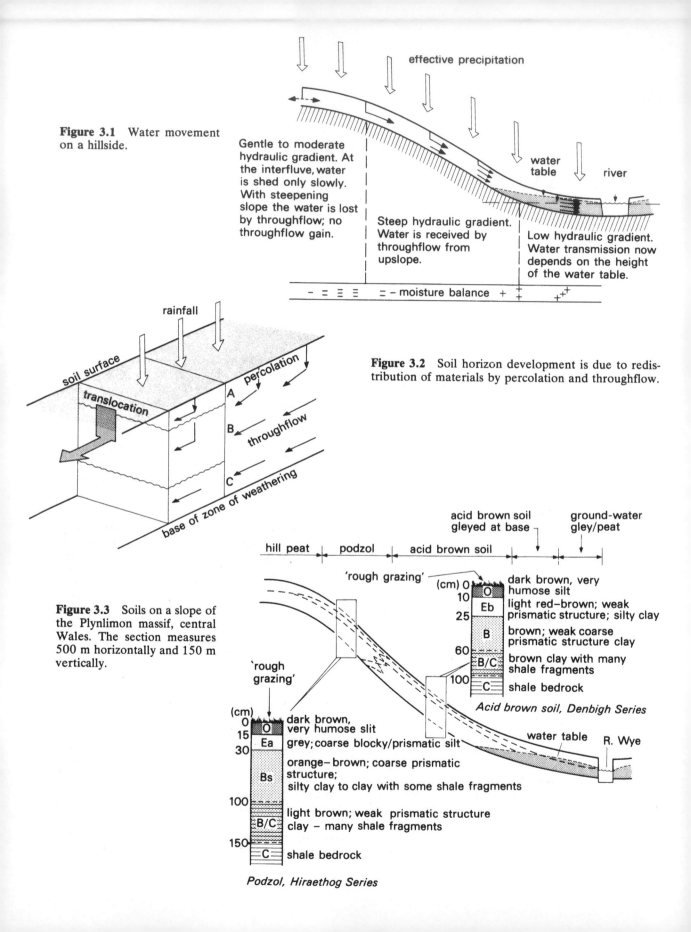

Figure 3.1 Water movement on a hillside.

effective precipitation

Gentle to moderate hydraulic gradient. At the interfluve, water is shed only slowly. With steepening slope the water is lost by throughflow; no throughflow gain.

Steep hydraulic gradient. Water is received by throughflow from upslope.

Low hydraulic gradient. Water transmission now depends on the height of the water table.

water table

river

− = ☰ ☰ ☰ − moisture balance +   +⁺ ⁺⁺

rainfall

soil surface

translocation

percolation

A

B   throughflow

C

base of zone of weathering

Figure 3.2 Soil horizon development is due to redistribution of materials by percolation and throughflow.

hill peat | podzol | acid brown soil

acid brown soil gleyed at base

ground-water gley/peat

'rough grazing'

(cm) 0
10
25
60
100

O — dark brown, very humose silt
Eb — light red–brown; weak prismatic structure; silty clay
B — brown; weak coarse prismatic structure clay
B/C — brown clay with many shale fragments
C — shale bedrock

Acid brown soil, Denbigh Series

Figure 3.3 Soils on a slope of the Plynlimon massif, central Wales. The section measures 500 m horizontally and 150 m vertically.

'rough grazing'

water table   R. Wye

(cm)
0
15
30

100

150

O — dark brown, very humose silt
Ea — grey; coarse blocky/prismatic silt
Bs — orange–brown; coarse prismatic structure; silty clay to clay with some shale fragments
B/C — light brown; weak prismatic structure clay – many shale fragments
C — shale bedrock

Podzol, Hiraethog Series

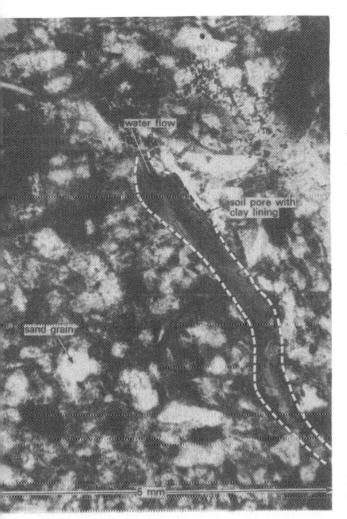

Figure 3.4 The movement and redeposition of clay particles in suspension are seen as dark linings to pores in this thin section.

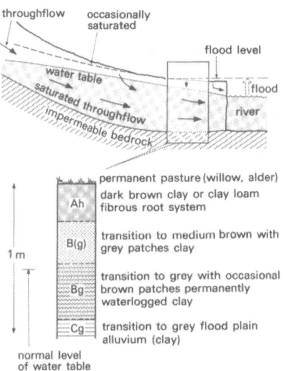

Figure 3.5 The position and profile of a ground-water gley soil.

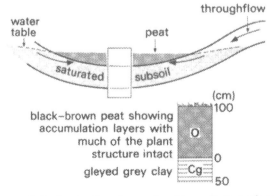

Figure 3.6 Peat often forms in depressions such as between drumlins and in kettle holes as well as beside some streams.

Total gleying thus reflects prolonged saturation whilst mottling is found in soil horizons only subject to saturation following storms.

In many flood plains the upper part of the soil is waterlogged for less than two months each year (about the time period needed for gley colour to become visible) and hence they are unaffected. The soil in this upper region is usually a dark brown colour, representing the combined colouring effects of the ferric iron compounds (orange) and organic matter (black). There is little downward movement of materials in flood plain soils because most flow is laterally to the river. As a result there is very little vertical translocation of clays and the soil profile has a fairly uniform texture.

In a ground-water gley, micro-organisms can usually survive in sufficient numbers in the upper

Figure 3.7 Extensive peat deposits, dissected by surface runoff, central Wales.

Figure 3.8 Surface water ponded above a surface-water gley soil.

parts of the profile to prevent the accumulation of undecomposed organic matter on the surface, but under extreme conditions, when water permanently saturates the whole profile, this is no longer possible. Two events then take place: first, the vegetation is reduced to those species that can survive waterlogging, such as mosses and reeds; secondly, as this vegetation dies it remains on the surface, gradually accumulating as a **peat** layer. Underneath this the mineral horizons are unchanged except that they become uniformly grey. Peats are found in very poorly drained parts of flood plains such as back channels (Fig. 3.6), by lakes, in enclosed depressions like kettle holes and on gently sloping interfluve areas in upland regions.

Areas of peat soils perform a unique hydrological function. The mass of dead peat acts like a giant sponge often causing the water table to rise above the original ground level. Water storage in this situation is very large and can result in large amounts of storm runoff being trapped. However the permeability of the peat is low so that, under wet conditions, when the water-holding capacity is filled, peat areas act like impermeable surfaces and promote surface runoff. It is for this reason that many peats contain their own channel-systems, eroded by surface runoff at times of heavy rainfall (Fig. 3.7).

The special effects of clay parent materials Many soils are developed in fine-grained materials deposited by the river as alluvium, while others are developed from clay, shale or slate bedrocks. On gentle slopes near to the river such soils will drain very slowly due to the small size of the soil pores and rainfall may remain on the surface for long periods. Soils in such situations develop gleying

features which are most pronounced at the top of the profile and may even be absent at depth. Most of these soils are classified as **surface-water gley** soils (Fig. 3.8) and are most commonly found away from the ground-water effects adjacent to rivers.

The Vale of the White Horse near Wantage The River Ock is one of several rivers that drain the claylands to the north of the chalk escarpment in Berkshire. It flows in a landscape of lowland relief, but even so ground-water influences are restricted to areas immediately next to the river channels and in the main, poor drainage is caused by low hydraulic gradients and clay textured soils (Fig. 3.9).

Nearest to the freely draining soils of the chalk scarp are the Ford End series soils. These are developed along a spring line and occur at the places where springs emerging at the clay/chalk junction give saturated conditions. Because of this the series has a discontinuous distribution. The soils are chalky and naturally permeable so they can be classed as ground-water gleys. (Note however that profile colours are not typical of gleys because there is very little iron in the chalk.)

Instead of basal gleying, as with Ford End soils, the heaviest gleying in Denchworth series soils is found near the surface, particularly in the Ag horizon. The cause of this is not only the low permeability associated with a clay texture but also because of a coarse prismatic structure in the B horizon. Water thus has few opportunities to use structural voids because they will be far apart (Fig. 3.10). A crumb structure, or at least a fine blocky structure would be needed here to improve the drainage. As it is, water cannot penetrate through the A horizon in sufficient quantity and surface gleying results. The gentle slope of the land also gives little opportunity for overland flow and so

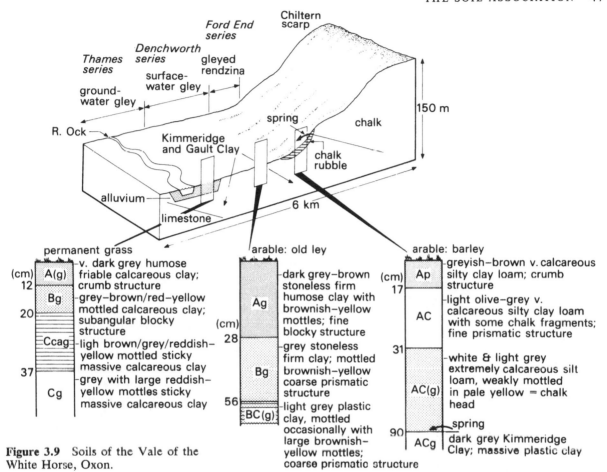

Figure 3.9 Soils of the Vale of the White Horse, Oxon.

water stands on the surface after heavy or prolonged rain. It is only nearer the river that the influence of a water table causes basal gleying and here a ground-water gley occurs again. Developed in alluvium this soil has a different profile from that of Ford End soils and is classified as Thames series.

In much of lowland Britain from Essex to the central valley of Scotland, glacial deposition is an important factor in landscape formation. The majority of deposition from an ice sheet is in the form of a thick layer of fine textured material (till) which masks the solid geology completely. The till layer varies considerably in thickness giving rise to widespread areas of 'confused topography', represented here by part of the drumlin field near Kilmarnock.

The Kilmarnock area   The soils map for this region shows a pattern of oval areas of one soil series surrounded by more sinuous bands of another series as the upstanding, and thus better drained, drumlins are separated from the intervening troughs (Fig. 3.11). However, because of fairly high effective rainfall (about 1500 mm per year) due to the northerly situation and clay textures, even the drumlin crests are only imperfectly drained. Nevertheless there remains a clear differentiation between drumlin and trough based on drainage, with the trough drainage classified as poor.

The soil association

The examples above illustrate that there is often a close relationship between hydrology and soil development in an area of uniform climate. This is particularly true for upland areas and has even been recognised as a basis for classification of soils by the Soil Survey of Scotland. The **soil association**

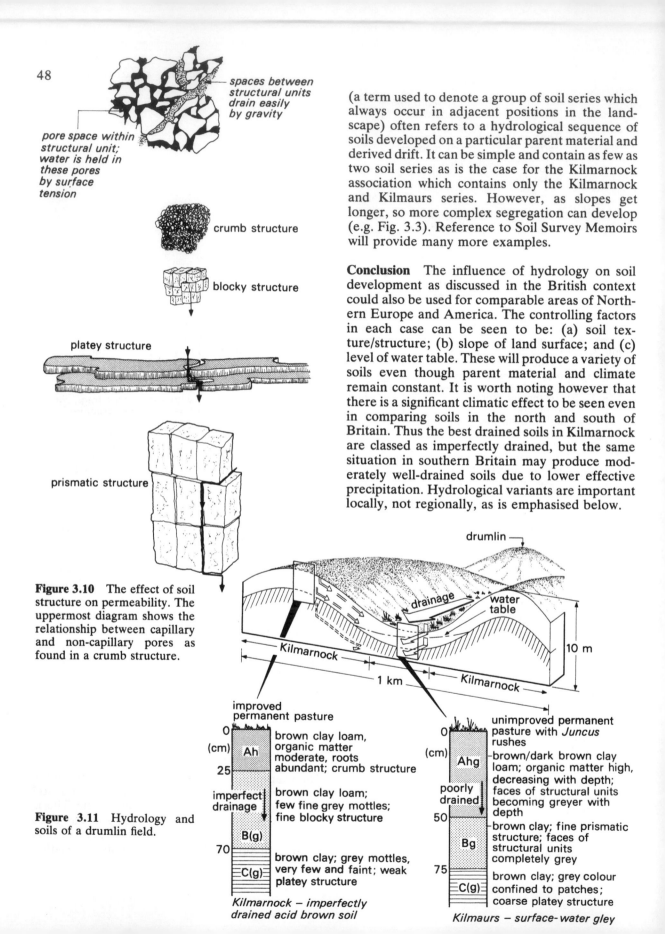

(a term used to denote a group of soil series which always occur in adjacent positions in the landscape) often refers to a hydrological sequence of soils developed on a particular parent material and derived drift. It can be simple and contain as few as two soil series as is the case for the Kilmarnock association which contains only the Kilmarnock and Kilmaurs series. However, as slopes get longer, so more complex segregation can develop (e.g. Fig. 3.3). Reference to Soil Survey Memoirs will provide many more examples.

Conclusion The influence of hydrology on soil development as discussed in the British context could also be used for comparable areas of Northern Europe and America. The controlling factors in each case can be seen to be: (a) soil texture/structure; (b) slope of land surface; and (c) level of water table. These will produce a variety of soils even though parent material and climate remain constant. It is worth noting however that there is a significant climatic effect to be seen even in comparing soils in the north and south of Britain. Thus the best drained soils in Kilmarnock are classed as imperfectly drained, but the same situation in southern Britain may produce moderately well-drained soils due to lower effective precipitation. Hydrological variants are important locally, not regionally, as is emphasised below.

spaces between structural units drain easily by gravity

pore space within structural unit; water is held in these pores by surface tension

crumb structure

blocky structure

platey structure

prismatic structure

Figure 3.10 The effect of soil structure on permeability. The uppermost diagram shows the relationship between capillary and non-capillary pores as found in a crumb structure.

Figure 3.11 Hydrology and soils of a drumlin field.

drumlin

drainage

water table

Kilmarnock

1 km

Kilmarnock

10 m

improved permanent pasture

0

(cm)   Ah

25

imperfect drainage

B(g)

70

C(g)

brown clay loam, organic matter moderate, roots abundant; crumb structure

brown clay loam; few fine grey mottles; fine blocky structure

brown clay; grey mottles, very few and faint; weak platey structure

Kilmarnock – imperfectly drained acid brown soil

unimproved permanent pasture with *Juncus* rushes

0

(cm)   Ahg

poorly drained

50

Bg

75

C(g)

brown/dark brown clay loam; organic matter high, decreasing with depth; faces of structural units becoming greyer with depth

brown clay; fine prismatic structure; faces of structural units completely grey

brown clay; grey colour confined to patches; coarse platey structure

Kilmaurs – surface-water gley

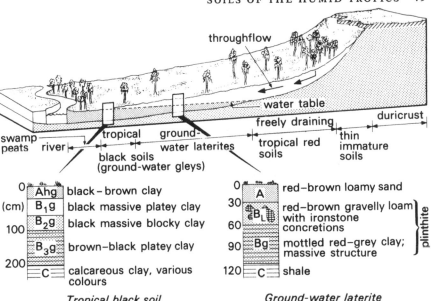

Figure 3.12 Soils developed in a tropical humid climate. Length of section about 8 km.

Tropical black soil   *Ground-water laterite*

Soils of the humid tropics

Over vast areas of tropical Africa and America there has been a period of climatic and tectonic stability lasting for hundreds of millions of years, so that landscapes have been reduced to widespread, gently sloping plains (peneplains). In such a landscape, throughflow will be slow even in humid areas where the high temperature and rainfall have combined to produce soils tens of metres deep.

This deep intense weathering can be expected to produce a wide range of soluble products whose movement laterally and vertically produces considerable soil variety. It is important to realise though, that, because of the uniformity of the landscape, changes in soils will be less frequent than in the humid temperate areas described earlier. A comparison of scales on Figures 3.12 and 3.9 will reveal the contrast and explain why there is a noticeable increase in the zone of influence of ground water in the humid tropics. Such zones fluctuate widely in area on a seasonal basis in sympathy, with the contrasting flow conditions (pp. 30–1).

Most freely draining slope soils are classified as **ferruginous** soils. In many ways they are similar to brown earth soils in temperate areas, but the intensity of weathering processes has resulted in red instead of brown as a dominant colour. Characteristically, the soils are deep, reflecting the efficiency of weathering under high-temperature conditions. All minerals are subject to chemical

reaction and although it appears that aluminium, iron oxides and silica are the most stable, some alumina and silica are still produced during leaching and they often combine in regions of slow flow to form clay minerals. It is for this reason that many tropical soils on low-lying plains have clay textures.

Many river margins have peat and ground-water gley soils in just the same way as in temperate areas. The real difference is that they spread farther on each side of the river because of the gentle gradient. However, in the wide zone where the soil-water table fluctuates with the seasonal rainfall **ground-water laterite** soils are found and these have no parallel in temperate regions.

As shown in Figure 3.12, these soils have a very distinctive horizon of iron and aluminium accumulation (B_L) which coincides with the region of water-table fluctuation. It is almost as though this zone acts as a fine filter, extracting soluble ferrous iron from the soil water during the wet season and fixing it in the insoluble ferric form when aerobic conditions prevail in the dry season. However, throughflow from upslope is also important here because the amount of iron in the profile is far too much to come only from other parts of the vertical profile.

It is interesting to note that changes in hydrological regime can alter these soils irreversibly, for example where streams have incised into the ground-water laterite exposing it to the air. Because of the incision the water table is permanently lowered and the B_L (**plinthite**) horizon dries out, eventually hardening irreversibly to a rock-like ironstone (**duricrust**) which used to be referred to as laterite.

Conclusions Hydrological factors in the humid tropics produce effects different from those in temperate areas largely because of contrasts in landscape. Gley soils and those on freely draining slopes are often of relatively little importance compared to areas occupied by the ground-water laterite and fossil soils (duricrusts). Many high level plateaux (like the Jos Plateau of Nigeria) have thousands of square kilometres of duricrust representing a change of hydrological conditions consequent on slight tectonic uplift of millions of years ago.

Soils of arid lands

In arid environments the lack of effective precipitation has resulted in vegetation and soils being generally absent. It is recognised that physical weathering can do little to produce a soil. In arid areas therefore the places to look for soil development are not the extensive hamadas or ergs, but where water can gain access to the surface on a permanent or semi-permanent basis. Such situations include oases, the valleys of exotic streams and playa lake basins. The least productive of these areas are the playa lake margins, as is clearly revealed by the fact that they are not cultivated by the indigenous population. Water reaching playa lakes comes from floods as described on p. 27, but as there is no outlet from a playa, water loss is primarily by evaporation, and the lake becomes increasingly saline.

Soils bordering such lakes are classified as **halomorphics**. Water seeps into the base of the soil profile from the lake and moves up to the surface by capillary action. At the surface any moisture is immediately evaporated, leaving the salts carried by the soil water as crusts on the surface (Fig. 3.13).

Oases are constantly fed by ground-water supplies and therefore remain of a stable size (Figs 3.14, 3.15). As with playas, water balance is negative except for a few hours during and following the rare storms; water therefore moves from the aquifer to the surface where it is evaporated. If the ground water has a high salt content, soils near oases remain saline and infertile whilst the sodium salts prohibit flocculation and the mineral particles remain dispersed.

Most water in arid environments comes from exotic streams that flow from mountains into the deserts where they dry up because of continued evaporation and seepage. Chemical weathering in arid environments is not very good at breaking down rock into small particles, so alluvial deposits tend to be of coarse grade and the soils invariably sandy. However, coarse-grade alluvial deposits make very good aquifers in which water travels quickly. Figure 3.16 shows the way in which water is lost from the river bed by seepage and stored in the alluvium. Capillary action brings this water within the plant root zone over most of the wadi floor. By way of contrast with playas and oases these supplies are not usually very saline as the supply is originally from rain or snowmelt in mountains.

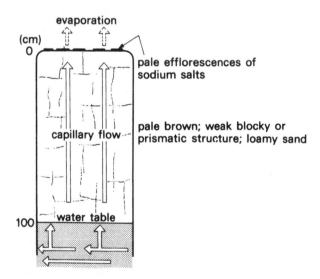

Figure 3.13 Desert soil development. The soil is called a solonchak.

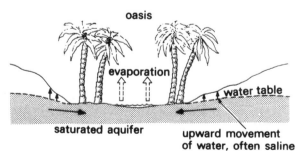

Figure 3.14 The hydrology of an oasis.

Figure 3.15 An oasis in the Moroccan Sahara. Note the absence of soil. Vegetation grows where water reaches near the surface.

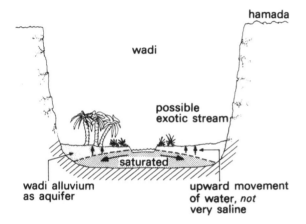

Figure 3.16 The hydrology of a wadi showing water movement without irrigation.

Conclusions Soils in arid situations are distinctive in that water moves upward in the profile. A low-rate of chemical weathering and the restricted availability of organic matter retards the development of soils considerably. The domination of saline minerals leads to soil dispersion and lack of structure.

The role of hydrology in soil management

Soil is one of man's most fundamental and important resources. It is used to support life by cultivation; it is used to build with, build on and drive over. It is vital to understand how a soil develops if it is to continue to perform the functions we have come to take almost for granted.

Plants use soil water as a source of basic foodstuffs in the form of soluble nutrient compounds. In addition they require oxygen for respiration and this they get partly from soil air. As plants use the same pore space to take up oxygen as well as water it follows that the soil must have some pores free

from water. Thus too much water is as inhibiting to growth as too little for the majority of plants. (There are some exceptions, such as rice.)

From a plant growth standpoint it is vital to maintain a soil moisture content somewhere between saturation and wilting levels. In practice, the best conditions for plant growth have been established as being near field capacity when most structural pores are air filled, but nearly all capillary pores are water filled. A freely draining soil having a crumb structure and loam texture would be expected to drain from saturation to this condition within forty-eight hours. Management for agriculture is thus designed to keep the soil as near to field capacity as possible, either by drainage or irrigation as appropriate.

Drainage systems British soils having drainage problems are much more common than those suffering from drought, so that soil hydrology management becomes a matter of promoting a situation as closely resembling free drainage as possible. With better drainage comes the prospect of flexibility in planting and harvesting times, the potential to grow a wider range of crops and make more extensive use of machinery on ground which is dry enough to withstand its use. But the adoption of drainage schemes rests with a satisfactory balance between cost and the possible benefits. For successful drainage any rainfall and throughflow must move through the soil easily so that air and nutrients can readily reach the plant root zone.

Control of the water table in a soil is fairly straightforward, using a system of clay pipes (**tiles**) laid in trenches 70–150 cm deep. These are placed in rows across a field with outflows to open ditches at field boundaries (Fig. 3.17). Each tile drain is

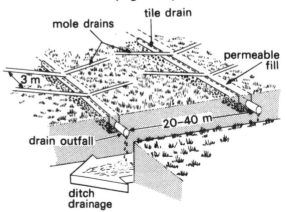

Figure 3.17 Field drainage arrangement.

Figure 3.18 Tile drainage in progress.

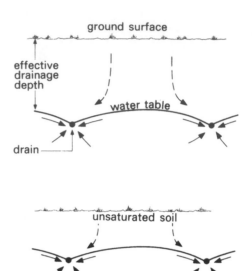

Figure 3.19 Water tables can be controlled at the time of drain installation by selecting the drain depth.

often covered with sand or another permeable material to 40–50 cm from the surface, while the rest of the trench is filled with soil (Fig. 3.18). Figure 3.19 indicates how the water table level is controlled by the pipe depth and spacing.

With fine-textured soils of low permeability, drains need to be spaced more closely, but a very close spacing of tile drains becomes uneconomic, so other systems have to be considered for use. In their natural state, fine-textured soils rely on structural fissures for water conduction as the spaces between individual particles becomes too small to be effective. There are no really satisfactory means of increasing the number of fissures permanently, but one technique often employed is called **subsoiling**. A square-sectioned plug is dragged behind a tractor at 40–50 cm depth (i.e. below normal plough depth) which lifts and breaks up the soil to make it more permeable. Subsoiling is a process which needs to be repeated every few years. Alternatively **mole drainage** is used (Fig. 3.17). Here a bullet-shaped plug is drawn through the soil at 40–50 cm creating a drainage pipe in the soil

behind it. The mole lines (usually 2·5–3·0 m apart) are arranged to intersect the tile drain permeable fill. The moles can be formed only in soils with structural strength and even then may need reforming every few years to maintain their effectiveness.

Because soil drainage is largely a matter of economics, areas have to be large enough to enable large-scale mechanised arable farming to function after improvement. Equally, on pasture lands the upgrading of grassland must provide an adequate return in terms of grass quality or prevention of diseases such as foot rot. Drainage systems are most effective when soil properties restrict permeability (i.e. their conditions leading to the formation of a surface-water gley). Drainage of areas with a high water table (giving ground-water gley soils) is less successful due to outfall problems and this is the reason why such schemes are not often found adjacent to rivers. However, in places like the Vale of the White Horse (Berkshire), there is much scope for artificial drainage. As discussed earlier, the major soil series (Denchworth) belongs to the surface-water gley group. The climate of the Vale permits arable cultivation, but because of the drainage difficulty the area used to be almost entirely permanent pasture. There have been fairly widespread attempts at land improvement by using tile drains at 75 cm depth with subsidiary moling at 55 cm. This is possible in all areas except near streams where some of the open ditches become flooded in winter. The careful planning to avoid such areas has resulted in a patchwork of arable (improved) land and pasture (unimproved with poor potential drainage conditions).

Note that the most extensive drainage improvements have taken place on surface-water gley soils. Ground-water gleys, such as the Thames series are, by their very nature, impossible to drain except by pumping on a large scale. In the Vale of the White Horse this is clearly uneconomic and so all that can be done is to keep field ditches clear so they may conduct floodwater away from fields as quickly as possible. Upgrading to arable use is therefore not possible.

In upland regions there are not only soils of low permeability but much higher effective rainfall totals than in lowland situations. A considerable amount of such land is owned by the Forestry Commission, including the eastern slopes of Plynlimon, Central Wales. Here, with an elevation of between 300 m and 700 m and an average rainfall of 2000 mm, most soils have a drainage problem. Tile drains are only of limited use and are mainly uneconomic, so ditches are much more freely employed on slopes as well as on bottomland. Figure 3.3 shows how lower slopes have developed gley or peat soils and, while special types of trees have been cultivated to grow in wetter conditions (e.g. Sitka spruce), there are limits to their tolerance and no conifers can withstand permanent saturation of the root zone. It is the policy of the Forestry Commission in these circumstances to dig ditches about 1 m wide and 60 cm deep in rows a few metres apart across the area to be planted. Material from the ditches is placed on the adjacent soil, thereby increasing its depth while the saplings are planted in the drier ridges thus formed (Fig. 3.20). Such schemes cover many hundreds of square kilometres and have been made more effective by the introduction of special ploughs which are pulled by very powerful crawler tractors.

Irrigation systems Arid area development varies greatly in different parts of the world. Most improvement schemes require very large capital resources – the most imposing are therefore to be found in such places as the Mid-West and West of the USA. Here soil moisture content is monitored and sprinkler irrigation applied to keep the soil at a specific moisture content just below field capacity. However, to give an impression of the scope available with more modest resources, the discussion below is concerned with management problems of

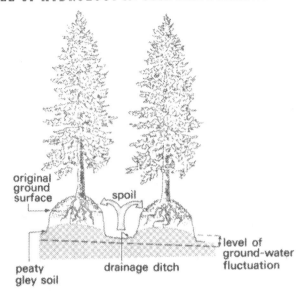

Figure 3.20 Ditch drainage to permit afforestation.

Figure 3.21 Irrigation in a wadi, Pont du Ziz, Morocco.

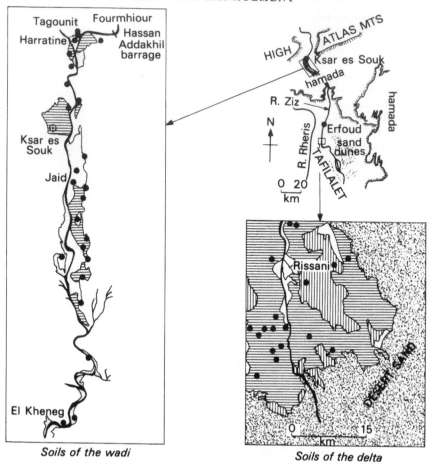

Soils of the wadi   *Soils of the delta*

▢ immature alluvial soils
▤ slightly evolved soils with profiles
▥ halomorphic soils with salty ground-water
• village

Figure 3.22 The pattern of soils in the Ksar es Souk area, Morocco.

the Ksar es Souk region of Southern Morocco. In the limited space available, this is offered as a total contrast to land drainage in Britain.

Ksar es Souk is a market town situation on the River Ziz, an exotic river draining southward from the High Atlas mountains to the Northern Sahara Desert. The river flows in a wadi down to Ksar es Souk (Fig. 3.21), whence it flows out on to a hamada depositing about two million tons of sediment a year to form a fan-like delta known as the Tafilalet. The alluvium deposited by the river is now over 7 m deep, providing a good depth of material for holding water and allowing soil development. The extent of the area affected by surface or ground water from the River Ziz is shown on Figure 3.22 and exactly corresponds

with the distribution of settlement. However, even though the river has a low salinity compared with aquifer supplies, the salt content is still of importance, especially in the southern delta. Between 1954 and 1964 the River Ziz carried an average of 112 000 tonnes of salt to the Tafilalet per year. Yet the streams that emerged from the southern part of the Tafilalet removed only 60 000 tonnes. The deficit of 52 000 tonnes of salt is a considerable restriction on soil development and agriculture.

Figure 3.23 shows a typical soil of the Ziz valley whose profile is everywhere alkaline with values of pH 7·5 to pH 8·4. Due to irrigation in the Ziz valley conditions are not as extreme as in uncultivated areas of the delta, where surface accumulation of salts is common. However, soils are very

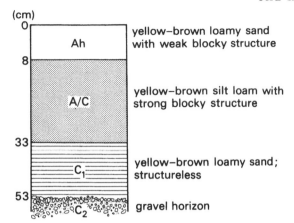

Figure 3.23 A desert soil after irrigation and agricultural use.

immature and have little organic matter content. The primary requirement for improvement is for a water movement from surface to depth to leach out the soluble salts. The scale of this undertaking is very large: merely to prevent a further build-up of salt in the soils of the Tafilalet, an extra three million cubic metres of water would have to be applied to the surface each year. Lowering the salt content to allow crops to grow would need much greater flows. In the southern Tafilalet 70 mg/l of salts have been measured in the ground water near the date palm groves. At present the groves remain unproductive because fruit will not form with salinity levels above 15 mg/l.

The problem is more than just one of supply. The water needed for intensive cultivation of all the Tafilalet soils is far greater than the supply which could be diverted from rivers. Nevertheless a great improvement is possible from the situation of the 1960s when only 4000–8000 ha of the total 19 000 ha were cultivated for cereal crops. In fact even the water that was then available could not be used to best advantage; with only primitive irrigation arrangements the fields were flooded only when there was water in the river. Water supply was merely a matter of natural storage in the alluvium until rainfall caused a rise in river level. Clearly with wasteful abstraction in the Ziz valley, relatively little was available to the Tafilalet.

To try to rationalise the situation, a comprehensive scheme to improve the agriculture of both the valley and delta was begun in 1968. A site a few kilometres upstream of Ksar es Souk was chosen for the construction of a 13 km-long reservoir, which would allow 280 000 000 m³ of water to be used for irrigation. The dam was completed in 1971, but a long period of filling is required when supply and demand are so finely balanced. It has been estimated that benefits to agriculture and increased production will pay for the dam and irrigation works in 40–50 years.

The reservoir allows a regular flow of water down the river within the limits of the rainfall. In consequence, fields are now watered twice a month with controlled amounts of water, thus allowing more intensive growth throughout the year. A subsidiary barrage at Erfoud is to control water to the Tafilalet. The problem of salinity has been tackled by the excavation of new drains: water is applied at the surface from feeder canals, percolates through the soil and the surplus is extracted via drains, together with soluble salts. This saline water is then returned to the river channel downslope of the Tafilalet and allowed to evaporate in the desert.

However, to make the most effective use of the water supply a new scheme of field arrangements is needed (Fig. 3.24). A large amount of tree cover is clearly necessary to protect the crops from intense sun and to reduce evaporation. Water

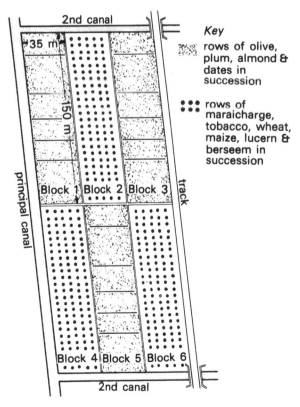

Figure 3.24 The pattern of newly irrigated areas.

control in this way still does not solve all problems. Unreliable rainfall will still reduce water availability for irrigation in some years, but an important improvement has been made.

Example problems

A. Discuss the effects of agricultural land drainage on: (a) soil development; (b) land surface erosion; and (c) river response to rainfall.

B. Figure 3.25 shows the variation in podzol profile morphology. Discuss the extent to which each may be attributed to changes in site hydrology or other factors such as texture, iron content of parent material and climate. Make sketches to indicate the positions in the landscape each profile might be expected to occupy.

C. Figure 3.26 shows the hillside profiles for soils above the River Exe floodplain. Why should the Halstow series be more gleyed than the Dunsford? Discuss, with the aid of diagrams showing probable through-flow/overland flow routes, the water flow pattern that might be expected from soil profile evidence and topography.

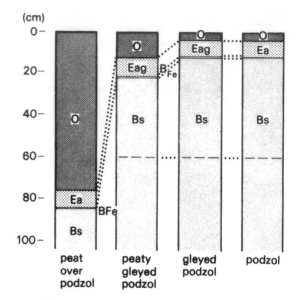

Figure 3.25 Variability in podzol soils.

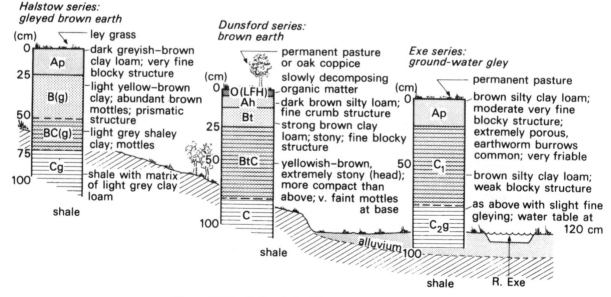

Figure 3.26 Soils of the River Exe valley, Devon.

Chapter 4

SLOPE DEVELOPMENT

Introduction

Geomorphological studies are increasingly concerned not just with landscape description but also with landscape development based on the identification and interpretation of causal processes. It is thus natural that hydrology and geomorphology should join forces in a study of landforms.

For any catchment within one climatic regime, we can assume that a single group of erosional processes are in operation, yet valleys show clear variations in cross profiles which suggest that the balance between processes has varied. The idea that a few variables can create many shapes is important because it enables us to consider widely different profiles as part of a continuum instead of as unrelated features. The formation of valleys can be attributed to (Fig. 4.1):

1. the rate of weathering of the bedrock;
2. the rate of transport of weathered material downslope;
3. the rate of removal of material delivered to the slope base;
4. the ability of the stream to: (a) carry material already in transit from upstream; (b) erode material from its banks and transport it; and (c) erode its bed (usually unweathered) and transport it.

In a short book such as this, only a glimpse of the range of hydrological influences can be given and, as processes within a stream channel are covered comprehensively elsewhere (e.g. Leopold, Wolman and Miller – see Further Reading), in this chapter we concentrate on the development of hillslopes.

Material is moved down a hillslope either as discrete particles, in solution and by overland flow, or in bulk, as in rockfalls, landslides and mudflows. A typical categorisation of such processes based on the movement type is given in Figure 4.2.

Rapid mass movement processes

Slides Many steep hillsides show evidence of instability. Indeed mudflows and landslides often cause considerable damage to property and are widely reported in the news. The result of these rapid mass movements is to leave a hillside temporarily scarred at the places where isolated sections have become unstable. In the long term, however, it is this action which has considerable influence on hillside evolution.

Clearly for movement to take place, the forces acting upon the soil or weathered rock must become unbalanced and, in the majority of such cases, such imbalance is the result of changes in the local hydrological conditions.

For example the weight of a mass of unsaturated soil imposes stresses on the material below it (**overburden pressure**, N) and downslope (**shear stress**, $S = mg \sin \beta$) (Fig. 4.3). Downslope movement is resisted by the internal sliding friction and cohesion of the soil (together called the **shear strength**, SS). **Cohesion** is a surface-tension effect produced by water films which acts to hold particles together while the soil remains unsaturated. As soon as the soil becomes saturated, the surface-tension effect is lost and cohesion declines to zero.

Instability and failure come about in two ways. The slope angle (β) may increase until the maximum value for dry shear strength is exceeded, as with a rock slide (debris avalanche). This works best when the cohesive force is naturally very small, as with coarse particles. Alternatively, and more commonly, the overburden pressure and thus the shear strength may be reduced by an increase in the soil moisture content until saturation occurs. As the depth of saturation (H_1) increases, so the pressure on water at depth increases and a positive **pore pressure** develops equivalent to an upthrust (μ) working against N. In the zone of saturation the cohesion is also zero so that the shear strength is reduced in two ways. It is possible

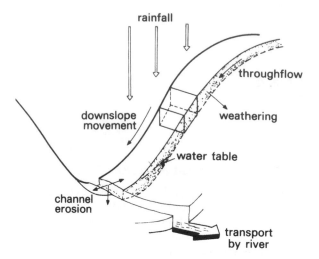

Figure 4.1 Processes involved in hillside and valley evolution.

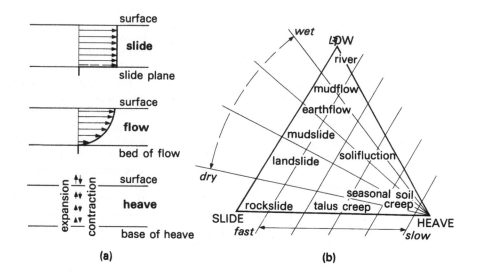

(a)

(b)

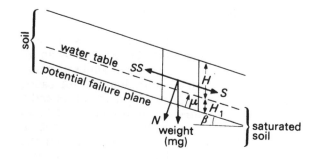

Figure 4.3 Forces acting on an element of hillside soil, part of which is saturated.

Figure 4.4 Shallow slides near Reading, Berks. The failure plane is clearly visible.

Figure 4.5 The 'Jackfield' shallow landslide of 1951–2, Ironbridge, Salop.

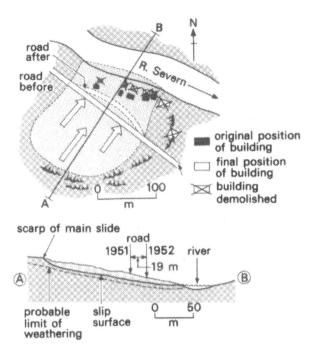

for the lower part of the soil to have zero cohesion whilst the upper part remains cohesive and this results in a tendency for the soil block to fail at depth along a shear plane and for the soil block to move as a unit. The ratio SS to S is called the **factor of safety** (F) for a hillslope, which must exceed 1·0 for continued stability. As geomorphologists are concerned with slopes on a long time scale, interest is focused on the maximum pore pressures likely to occur. The depth of saturation will increase during a severe storm if saturated throughflow builds up above a horizon of low permeability. The limiting case occurs when the water table rises to the ground surface and throughflow is parallel to the ground slope. As a result most slopes with cohesive materials (soils, weathered shale, clay, etc.) tend to be in equilibrium with the *saturated* shear strength.

Once initiated, slides continue to move as long as saturated conditions prevail. They come to rest either because of movement on to more gentle slopes (when shear stress decreases), or because water drains from the sliding mass by throughflow (thereby increasing the shear strength), or through a combination of both. When drainage is the main arresting factor, the slides occur in a series of intermittent steps, each step taking place during the period of saturated storm throughflow. The slides discussed above are called **shallow slides**

(translational slides) and rarely exceed a few metres in depth, although they may be hundreds of square metres in area (Fig. 4.4). They are probably the most common and certainly the most important mass movement process involved in the long-term evolution of steep valley slopes that retain a soil and vegetation cover.

The slide at Jackfield in the Ironbridge Gorge, Salop is fairly typical (Fig. 4.5), the 10° slope of the valley being too steep for stability of the weathered shale bedrock. This slide moved slowly and intermittently over a period of a couple of years, eventually resulting in a 20 m displacement of the road which crossed it and also the demolition of some houses. On investigation the slide proved to be totally in weathered shale with an average depth of only 6 m. Significantly the main movements occurred in the winter periods of 1951–2 and 1952–3 when, with water tables reaching the surface in a number of places, saturated conditions resulted in a reduction in shear strength. The slide moved intermittently because saturation prevailed only for the limited period after each storm before throughflow reduced the water content below its critical value.

Rotational slips Larger slides, usually called rotational slips, are also dramatically affected by hydrological conditions. Here we are usually

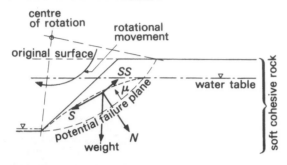

Figure 4.6 A long slope in soft rock provides the possibility of a deep-seated rotational slip.

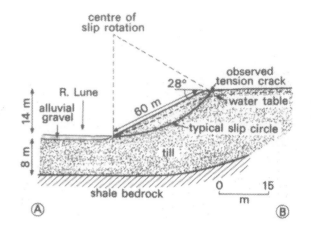

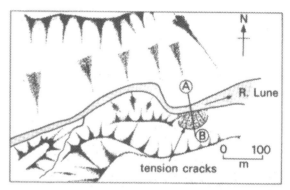

Figure 4.7 The 'Selset' rotational slip near Middleton in Teesdale, Yorks. This slip (shown stippled) moves very slowly. Historical records show the site as a river cliff for over a hundred years. Other parts of the hillside, not subject to undercutting, experience shallow slides.

Figure 4.8 Terracettes (foreground) and shallow slides, central Wales.

dealing with a slide in unweathered material such as a clay or glacial till, as well as the weathered surface layer. The same principles causing movement apply, but the situation is different (Fig. 4.6).

Figure 4.7 shows a section of the Lune valley in Yorkshire cut in till. Here the change from a surface slide to a deep-seated rotational slip is partly caused by the greater length of the slope. On a larger slope the shear stress on any particular failure plane is much larger because it increases directly with the weight of the soil above the plane; the increase in soil strength is much less because the cohesion value does not increase and only internal friction increases. The failure plane will clearly be helped by the presence of water, reducing the normal stress as before.

Shallow slides and rotational slips are not mutually exclusive and in fact usually occur together – the shallow slides taking place in the weathered mantle of the rock in which the deep-seated slide occurs. It is also believed that **terracettes** are part of this system, representing rotational slipping of the soil when it is too thin to slide (Fig. 4.8). Note that all these movements are to be expected particularly in regions of concentration of throughflow such as shallow concavities (p. 11).

In some extreme examples, the pressure of water in the pores of material causes it to exceed the liquid limit and the material then ceases to behave as a solid, and flows. Such occurrences are called **earthflows** or **mudflows** depending on the proportion of water to soil.

Translational or rotational slides can often develop into rapid flows. They are especially common in areas with steep slopes and unconsolidated bedrock where heavy rainfall can cause the development of positive pore pressures throughout the weathered mantle, especially when a network of subsurface pipes (p. 41) is present

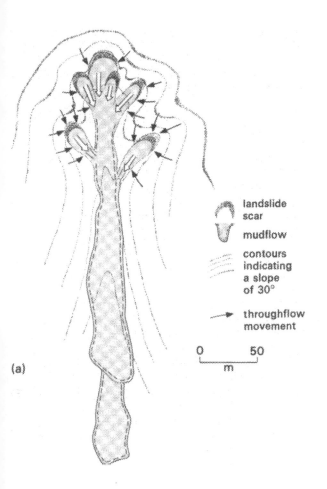

(a)

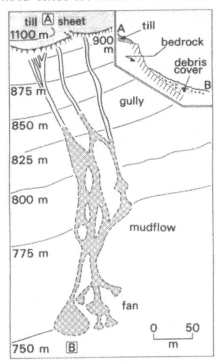

(b)

Figure 4.9 (a) Landslides can develop into mudflows in valleys with steep slopes developed in unconsolidated materials. (b) Mudflows without slides developed in a valley in northern Sweden after water saturated the till at the top of the slope. The mudflow cut gullies in the existing debris cover and deposited a fan at the slope base.

(Fig. 4.9 (a)). In such cases there is often no region of preferential failure, the whole mass becoming fluid at the same time. Very heavy rainfall often triggers catastrophic mudflows such as when 178 mm of rain in twenty-four hours caused 48 000 m³ of unconsolidated sandstone to develop many landslides which then formed into a complex mudflow and partially filled up a valley near Buzău in the Romanian Carpathians. However, long periods of less intense rainfall have also caused repeated but smaller mudflows in the same area. 21 500 m³ of debris is reported moved by mudflows over an eight-year period from one hillside alone. Unconsolidated till is also prone to flow as illustrated in the Karkevagge valley, Sweden (Fig. 4.9(b)) whilst there are even several recorded mudflows from steep vegetated hillsides in Britain.

Solifluction Periglacial conditions still cover one-fifth of the land surface and were probably even more extensive during the Quaternary cold periods. As such they are of special importance to geomorphologists. **Solifluction** is a term generally used to describe the slow flow of unconsolidated material saturated by water downslope during the summer period of a periglacial climate, and is equivalent to a slow earthflow in temperate regions. It is thought that solifluction is often connected with late lying snowpatches (Fig. 4.10). In such situations snow still melting through the late summer produces near saturated throughflow in the 'soil' downslope of the snowpatch right up to the time of freezing in winter.

Solifluction lobes and terraces can be seen as independent features on the hillside up to one hundred metres long, each downslope of a snowpatch of variable extent (Fig. 4.11). Both lobes and terraces would have ice in most of the pore space of the seasonally frozen layer throughout the winter and, with a thaw of the surface in early summer, the ice in the pores melts. Below this (about 50 cm) freezing conditions prevail much longer and so removal of water by percolation is inhibited. As a result saturated conditions occur in at least part of the seasonally frozen layer and pore water pressure counteracts overburden pressure allowing downslope slow flow or solifluction. However, within a few weeks of the start of movement, surplus water has drained out and the flow during the rest of the thawed season slows to that appropriate to **soil creep**. Locally, rates of movement are high – in this example up to 43 mm downslope per year as compared with 1 mm per year for a nearby slope only affected by soil creep.

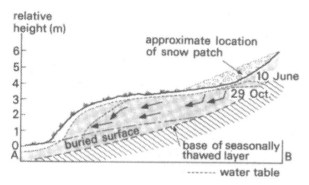

Figure 4.10 Section through a solifluction lobe. Note the throughflow pattern and region of saturation.

Figure 4.11 The front of a solifluction lobe, Okstind-bredal, Norway.

Surface-water erosion

At times of heavy rainfall it is common to see raindrops hit roads and other hard surfaces and splash up sometimes several centimetres into the air. If such raindrops fall on bare soil, part of the energy is given to the soil particles which are thrown up into the air with the splashing rainfall. On a slope there will be a tendency for most of the splash to occur in a downslope direction, so causing surface erosion. Notice that, in contrast to mass movements considered above, rainsplash is not of localised occurrence on a hillslope but affects it all for the duration of the high intensity rainfall. Measurements have revealed that several tonnes per hectare of soil can be continually in suspension because of rainsplash.

Under conditions where rainsplash is effective it is also likely that rainfall intensity will be in excess of infiltration capacity and so overland flow occurs. Initially this will be as a sheetflow, whereby the water travels over the hillside as a sheet of laminar flow. However the irregularities of the slope will soon concentrate the surface wash into depressions, creating concentrated turbulent flow which moves down the hillside gaining in speed until it has sufficient energy to entrain particles of soil. Erosion along these selected routes leads directly to the development of channels called **rills**. Rills tend to be shallow and of no fixed path, chang-

ing within storms and causing fairly even erosion of the upper hillslope. However, lower down they merge together to give more major flows which erode permanent channels called **gullies**. Gullies are most commonly found in semi-arid areas where discontinuous vegetation cover and high intensity rainstorms are the rule, but they can be promoted in other regions of steep slopes by the removal of vegetation for agricultural reasons. Note, however, that surface erosion by rainsplash can also be important in the humid tropics under rainforest. This is because the dense tree canopy prevents light from reaching the ground and thus prevents a vegetative cover. Drips falling from even drizzle in the forest canopy fall so far that they attain near to their terminal velocity and so can cause significant removal of soil on slopes.

It will be clear that all of these surface-water-erosion processes are inhibited by the presence of a continuous vegetation cover and cannot be expected to be important in temperate humid areas. The classic approach to erosion by surface water was conducted by Horton in South-West USA, where erosion rates have been recorded up to 10^4 cm^3/cm/yr. By contrast, for steep vegetated slopes in Britain the average rate of surface erosion seems to be of the order of 0·8 cm^3/cm/yr, a figure more than ten times smaller than soil creep in such areas.

In humid temperate areas the only place where significant erosion can be attained by surface water is in regions of frequent saturation by throughflow such as the concavities described on p. 10.

Subsurface-water erosion

Whilst overland flow is important in causing erosion under the conditions described above, many slopes dispose of their rainfall almost entirely by throughflow or percolation to aquifers as described in Chapter 1. Water moves through soil pores slowly and is not thought able to move soil particles more than locally, as in the formation of soil B horizons. Only when pores become enlarged on steep slopes to form pipes (p. 41) can we ascribe an important role to direct mechanical subsurface erosion. Most soil formation is the result of chemical reactions between bedrock and water, with the soluble products carried away as a direct loss to the hillslope.

The amount of material taken into solution will depend on many factors including: (i) water temperature; (ii) the length of time water is in contact

with the material; and (iii) the presence and nature of acids derived from soil organic horizons and atmospheric contamination.

Because chemical erosion is so dependent on temperature as well as moisture, it has varying rates of action over the Earth's surface. Nevertheless it is widely regarded as the most important agency of erosion in humid equatorial regions, of major significance in temperate regions and a still important role in the polar regions (Table 4.1). On a world scale it has been calculated that solution is the most important process for up to 35% of the land area (Table 4.2).

The nature of chemical weathering is not really influenced by climate, but the rate of resultant solutional erosion is. Clearly a hot wet environment

Table 4.1 Denudation processes in the Karkevagge Valley, Sweden (1952–60).

Process	ton/km^2/yr	ton m/yr
rockfalls	9	19 000
avalanches	15	22 000
rotational slips	20	75
sheet slides	23	20 000
slides/mudflows	26	76 000
creep	—	5 000
solifluction	—	20 000
solution	26	136 000

Table 4.2 Dissolved and solid transport of major rivers. (Note that atmospheric salt content of rivers is, on average, about 20% of the dissolved load.)

	Dissolved transport (ton/km^2/yr)	Solid transport (ton/km^2/yr)	Examples
mountainous area, high precipitation	100–500	200–1500	Brahmaputra, Ucayali, Magdalena, Rhone
mountainous area, low precipitation average relief	10–80	100–1000	Colorado, Amu-Daria
temperate or tropical climate low relief, dry climate	40–100	40–200	Danube, Mississippi, Parana, Madeira
	3–10	10–100	Chari, Murray
low relief, temperate climate	20–80	20–50	Rhine, Northern Divina, Volga
low relief, subarctic climate	10–40	1·5–15	Lena, St Lawrence, Finland
low relief, tropical climate	3–20	1–10	Congo, Negro, Tapajos

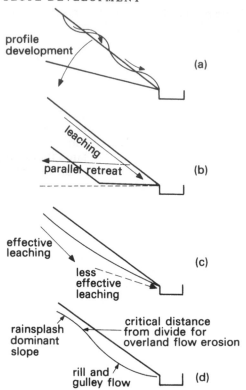

Figure 4.12 The evolution of slopes by single erosion processes.

will not only speed up reactions but there will also be sufficient throughflow to ensure the removal of the products. Under such conditions the whole slope loses material and there is a tendency for slopes to retreat in parallel. But when this is not so, redeposition in the lower part of the slope would tend to give greater net erosion to the upper slope section and a concave profile would tend to form.

Hillslope models

The considerable variety of hillslope erosion processes that result from changes in hydrological conditions prompts the question of resultant slope forms. Certainly each process, if left to continue without interruption and without changes of local base level, would probably lead to the development of a characteristic hillslope form (Fig. 4.12). Mass movements are particularly important in the deeply weathered soils of the humid tropics, but solutional processes are also important and would tend to lead to a different hillslope form. The same

conflicts arise within other climatic regimes so that a multifaceted slope is usually the rule, with each slope section a response to the dominance of one process.

We can envisage most slopes as having rounded convex summits, straight central regions and basal sections that are gently concave. The convexity may be due to either soil creep or rainsplash (according to the nature of the vegetation cover); the central straight section is probably a reflection of mass movements and the concave base is due to surface wash and solution. Each section will occupy a proportion of the slope according to the efficiency of the processes operating, such that at times there may be very little convexity, straight central region or even basal concavity.

Most early stages of humid climate river systems have steep valley slopes where vertical erosion is rapid. In these situations it is probably best to think of an initial phase with marked dominance of slides and mudflows, whose location is dependent on hillside throughflow patterns. Eventually slopes become too gentle for this to operate effectively and then creep, wash and solution take over to shape the slopes at a much slower rate. In less humid situations, where rainstorms are of high intensity and vegetation does not completely cover the slopes, wash is an important factor at a much earlier stage. Finally it is important to remember that some of the most important processes have an irregular movement, often with a frequency of only once in a hundred years or less, and this must be set against the millions of years valleys take to evolve.

A detailed slope analysis

Finally all the above mentioned relationships between hydrological status and slope development can be clearly illustrated with results from a section of escarpment near Rockingham, Northants. The oolitic scarp (here made up mostly of Upper Lias Clay with a capping of till) has been subject to much erosion (Fig. 4.13). Rotational slips, shallow slides and flow movements have all been at work in close proximity, although none appeared currently very active (Fig. 4.14). Rotational slipping has taken place on the steepest section of the slope (with the slip plane at the till/clay interface) whilst below it, on the 9° slope, a series of terraces has been created by a complex sequence of shallow sliding movements totally within the weathered clay. In the toe slope region there is a further sequence of terraces produced by slides and in this

Table 4.3 Material stability at Rockingham.

Position	Water table depth (m)	Safety factor (F)
upper slope rotational slip	estimated	1·28
upper slope slides	0·2	1·00
toe slope slides	0·8	1·32
	0·0	1·08

case a soil horizon has been buried. Investigation of the site hydrology revealed a water table on the slope mostly between 0·2 m depth (winter) and 1·6 m (summer). Taking the maximum figures and using them in an equation for weathered clay stability derived from Figures 4.3 and 4.6 gave the results in Table 4.3. In all cases there is stability at present, even under the highest water table conditions encountered during the measurement period. A safety factor below 1·0 can only be obtained by increasing the water content further such as would occur during heavy and prolonged storms. However, old maps show no evidence of slipping in the upper slope for at least the last 300 years, while a spring prevents the build up of high pore water pressures near the rotational slip. Pore water pressure could only build up if the spring were blocked, for example in periglacial conditions by freezing. Whilst the upper slope terraces show a safety factor of 1·0, indicating that they could move at present if the soil were bare, such a condition is likely only in cold conditions. The terraces are also consistent across the slope and show some signs of flow structures so they may be likened to solifluction terraces and again probably date from the last glacial period when a snowbank may have existed below the scarp crest.

The lower slope terraces are pure slides and Table 4.3 suggests that they would be very near limiting equilibrium if the water table could reach the ground surface. Some of these slides are very old, as is indicated by a buried soil horizon dated at 2200 BP but there has clearly been movement since, firstly to bury the soil and secondly to disturb the pattern of medieval ridge and furrow which was established on the toe slope.

In the middle of the section AB there is a knoll of undisturbed clay. This may well be the remains

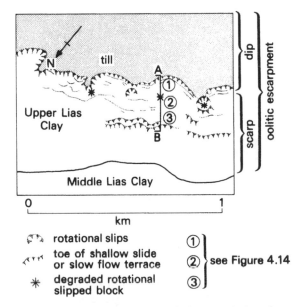

Figure 4.13 The morphological characteristics of part of the oolitic escarpment near Rockingham.

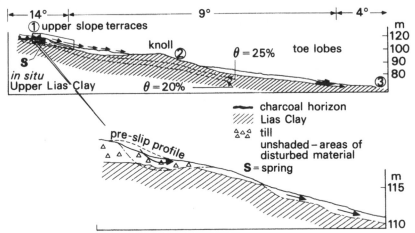

Figure 4.14 Cross-section along the line AB of Figure 4.13.

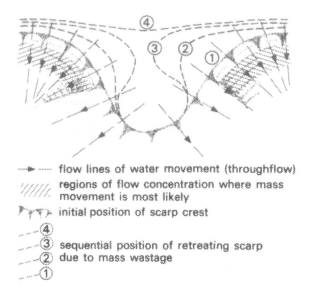

—→ ---- flow lines of water movement (throughflow)

///// regions of flow concentration where mass
       movement is most likely

▸▾▾▸ initial position of scarp crest

--④
--③ sequential position of retreating scarp
--② due to mass wastage
--①

Figure 4.15 The development of a small scarp face
knoll by mass wasting.

Figure 4.16   A steep hillside, Plynlimon, central Wales.

of a small scarp face promontory, as indicated
in Figure 4.15. Note how patterns of water
movement in such situations leave the end of the
promontory with a low moisture status, whilst con-
centrating water along the sides. These side
regions would then become liable to increased
mass movements and the end of the promontory
would eventually become isolated from the scarp
without any deep-seated movement or surface
stream action.

The scarp near Rockingham illustrates how a
hillside can be subject to a variety of mass movement processes whose frequency of operation is
controlled by the hydrologic regime over a long
period of time. It is such detailed investigation that
will allow a better understanding of all aspects of
geomorphology.

An example for discussion

Figure 4.16 shows a steep hillslope from the Plynlimon
region, Central Wales. Using data from pages 12–14,
17–19, 43–47 and the concepts discussed in this chapter,
discuss the importance of rapid mass movement processes
in such a situation. Attempt to explain the significance of
the 'flushes' seen on the slope.

Chapter 5

WATER RESOURCES

The scope of the problem

'This morning the water industry seems to be fairly good: the tap works; the sewage disappears, and if one wants to go fishing one can get it somewhere. So why is everyone worried?' (Sir Norman Rowntree, as Director of the Water Resources Board, 1974).

'There is no intrinsic shortage of water in England and Wales. In total rainfall is ample to meet all demands for the forseeable future. The problem arises from the uneven distribution of available water over both time and place' (Water Resources Board, 1973).

As the first statement indicates, the main problem is that, in general, we take water supply for granted. Only when there is too much or too little does it seem really important. And whilst most of us accept water supply without much thought there is, in fact, a whole industry whose sole purpose is to make sure that water is supplied in the right quantity, at the right quality and at the right place at the right time. There are thousands of people employed in the water supply industry and tens of thousands more involved in supplying its equipment, from dams to stop cocks, from raingauges to computers. So here hydrology and man interact in the provision of water for irrigation, industry and domestic use, for amenity, transport and so on.

Without man's intervention in the hydrological cycle, the only supply that could be relied on would be the dry weather flow, which occurs naturally after a long spell of dry weather (Fig. 5.1). This is only of the order of 10% of the average river flow and in some lowland clay catchments may be even smaller. Of course such a flow would only provide a small part of the water needed to satisfy all our present demands. Man is therefore compelled to interfere, his effectiveness depending upon his understanding of the cycle, his technological abil-

ity and economic considerations. It is, for example, possible to raise the dry weather flow from its natural 20 000 000 m³ a day to 32 000 000 m³ a day in England and Wales by storing only 2% of the annual water flow, and this is at present what our surface reservoirs provide. The cost, although large, is justifiable and the construction of suitable reservoirs not a very difficult problem. However, it is not economic to guard against the sort of drought that occurs once every fifty or more years because the storage capacity needed would be very large.

For most of the year, the concern is with supplying enough water at a constant rate, but from time to time heavy rainfall produces too much water, and rivers flood, so that a degree of interference is needed here too. And whilst regulating supply, careful thought must also be given to the maintenance of river quality so that the natural flora and fauna may thrive, water bodies may be used for public supply, leisure, transport and to enhance the landscape.

The integration of all these aims needs careful planning in a way that is as flexible as possible so that, as demands change, there can be smooth adaptation at minimum cost. These are not small decisions: reservoirs, pipelines and aqueducts built today may have a lifespan of over two hundred years; the costs involved are tremendous and their impact on the environment substantial.

Water supply

Public water supply, that is clean drinking (potable) water delivered via a treatment plant, can come from: (a) rivers, (b) aquifers, (c) reservoirs, and (d) the sea (via desalination). **River water** provides the most easily accessible source, but the quantity abstracted is often limited because, in order to maintain water quality, a certain amount of water has to be kept in the river to allow for effluent dilution. Because of this the amount of

Figure 5.1 The dry weather flow of the River Wye, Plynlimon. At times of high flow the water can fill and overtop the measurement flume.

water that may be abstracted decreases in a period of drought at just the time when the demand increases. Industrial users obtaining water direct from the river face the same problem and often public water supply and industry find themselves in direct competition for the same river water. Clearly this problem becomes greater still as a catchment becomes more populated and so some form of storage has to be provided to guard against times of low flow. It would be very useful to use natural lakes to store water, but this is rarely done because the changes in storage required entail changes of lake level and many metres of muddy shore exposed in the summer. There is therefore a direct conflict with landscape and recreational uses, a conflict which water supply usually loses. For example, few of the Lake District lakes have been adapted for water supply purposes.

In many other areas there are no lakes and so the general solution has been to construct artificial **reservoirs**. The older reservoirs, such as those in the Elan valley in central Wales, were constructed solely for the purpose of supplying clean water. They are therefore found in upland areas where unpolluted water can be stored and transferred directly to urban areas. But because unpolluted water is to be transferred, it is not possible to allow public access to these reservoirs for recreational

purposes. However, the pipelines needed for this transfer are expensive to construct and maintain, whilst better water treatment is now available than was the case when these early reservoirs were built. The problem now is to transfer water as economically as possible without too much regard to the protection of water quality. The cheapest means of transport is to use the natural river network as much as possible and this can be achieved by regulating the river flow with a different design of reservoir. Such a **regulated river system** has a lot of other benefits as well. There is no longer a need to keep the public away from the reservoirs because water is treated at the river abstraction point to a standard necessary for human consumption. However the most important advantages are concerned with flood control (p. 70) and provision for increasing demand. When a reservoir simply regulates the flow of a river, abstractions can be made at the most convenient places. If these demands tend to cause the flow to become too small for the maintenance of a good quality, then more water is released to compensate. Each abstractor uses the water and then returns it, if possible, back to the river so that the water may be abstracted again downstream. The total release of water by the reservoir is therefore not the gross consumption of all abstractors, but the *net* consumption, which is far lower. In this way a reservoir can supply several conurbations without the need for enormous storage as would be the case if all water were being supplied directly from reservoirs. One of the most recent regulating reservoirs is on a tributary of the River Severn in central Wales near Llanidloes (Fig. 5.6).

Regulation of the River Severn is achieved by releasing water from Llyn Clywedog into the Afon Clywedog to sustain river flows whenever the flow at Bewdley Control Point in the Midlands is forecast to fall below the prescribed flow of 8.4 cumecs.

Forecasting of flows requires knowledge of the factors which affect river flows throughout the catchment area to the Control Point (4330 km²), such as abstractions from the river, sluice operations, increases in flow due to rainfall, decreases in flow due to groundwater depletion and reservoir releases. The time of travel of releases to the Control Point is three and a half to four days, so all these factors need to be assessed for expected behaviour during the next four days. For example: 'There has been no rain for a week and the rivers are running lower each day.' The dry weather causes a rise in the demand for public water supply, so abstractions from the river are expected to increase. As the flow in the Severn at Bewdley is falling at a rate which is calculated to produce a deficit below the statutory flow in six days, close contact is kept with the

Meteorological Office to see if rain is expected in sufficient quantity to increase river flows. No rain is forecast and hot weather sets in, increasing both evapotranspiration and demand for abstracted water. A release is authorised from Llyn Clywedog to arrive at Bewdley in four days' time to prevent the forecast deficit occurring. Performance of the release is monitored as it passes down-river and adjustments are made during succeeding days to counter further developing flow deficits (Severn/Trent Water Authority)

Despite the advances that have been made by the use of regulating reservoirs, an acute problem remains in catchments like the Thames that have no direct access to upland storage. In these cases the additional demands of lowland conurbations can be met by **pumped storage** reservoirs, such as those surrounding London (Fig. 5.2), which obtain water from the nearby river at times of high flow and then provide supplies direct to treatment plants. Although their role is vital, they are very expensive, taking up valuable space near to urban areas, and they exercise no control over river flow. Because of the large amount of land being occupied by pumped storage reservoirs, attention has been focused recently on the possible storage of fresh water behind barrages and embankments within **estuaries** such as the Wash (Fig. 5.3). In

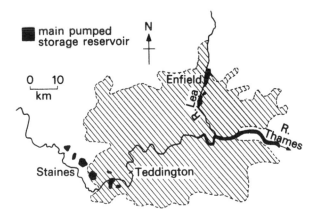

Figure 5.2 The major pumped storage reservoirs supplying London.

Figure 5.3 The Wash water-storage scheme. Early proposals for a four-storage bunded system have been revised based on new demand forecasts to a two-stage scheme. Stage 1, needed after 2001, would be built first with the possibility of adding stage 2 much later on.

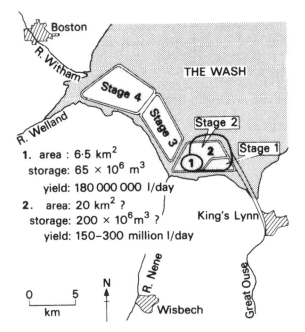

regions where reservoirs would have to be enclosed by artificial embankments anyway, these 'bunded' reservoirs in estuaries are quite economic.

In some areas large amounts of water are held in the natural storage of **aquifers**. Workable aquifers include the chalk, oolitic limestone and Triassic sandstone, and from these water can be pumped and used to supplement the public supply. Aquifers provide water of high quality, which therefore needs very little treatment, and they do not compete with other uses of the land surface. Despite the complex bore holes and well-pumping equipment needed, this supply is cheap and maintenance-free. There is a great temptation to overexploit this resource because of its convenience and there are many examples of overpumping causing water tables to fall many metres below their natural levels. Aquifers recharge slowly and therefore cannot be pumped too severely if they are to be of long term use. Many rivers are partly fed by aquifers so that, if water tables fall below river bed levels, rivers are either reduced in flow or stop altogether. There is thus little point in increasing yield by pumping if this results in a decreased river yield. Recently there have been attempts to pump some water back into aquifers when there is a surplus from river sources, but it is a very slow process. It is hoped that this artificial recharge might partly compensate for abstractions during drought (Fig. 5.4).

Only rarely can **desalination** of sea water be considered as an economic alternative to other sources of supply. At present desalination plants are very energy-intensive and as the cost of energy supply increases, without changed techniques the likelihood of desalination being commonplace in any but arid regions is remote.

Flood protection

Floods are a consequence of natural water exceeding the capacity of removal by the river channel and can be expected to occur at least once every three years. Such floods are important insofar as they provide the alluvium to flood plains which in the long term maintains their agricultural potential. In arid regions they are also vital in that they flush salts from soil profiles. Unfortunately such productive regions are also places where most people choose to live and work and so it is desirable to be able to offer some protection against the effects of flooding. One of the oldest systems of flood protection employed has been to construct **embankments** (dykes) along the river banks to contain the flood water in the channel. For instance there are 350 km of dykes along the lower Rhein in Germany alone. The Embankment in London is, in fact, a dyke system to contain the Thames. However the costs of this are so high that many less densely populated areas have had no real flood protection. One solution, which has the advantage of also helping water supply, is to try to contain at least part of the flood at source by means of a reservoir, releasing the stored water in a controlled way when the flood risk has passed. The Clywedog again provides an example of a regulating reservoir used for flood control.

Traditionally, people from Llanidloes to Shrewsbury could expect the River Severn to flood over its banks and into their streets at least once every three years, but a flood alleviation scheme was uneconomic. However, by the 1950s it was clear that abstractions from the River Severn were not going to be sufficient to supply new industry and centres of population in the Midlands. In summer the natural flow of the Severn was often little more than that needed to dilute effluent (Fig. 5.5). This

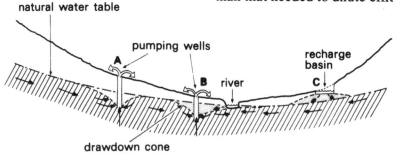

Figure 5.4 The utilisation of ground-water resources. Pumping at **A** lowers the water table but has little other environmental effect. At **B** pumping drawdown depresses the water table below river level and causes abstraction from the river. The water table can be raised locally as at **C**. Such recharge basins are mostly sited near to pumping stations so that some of this water can be recovered in times of need.

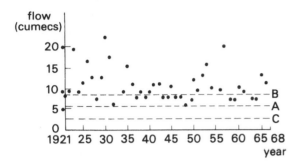

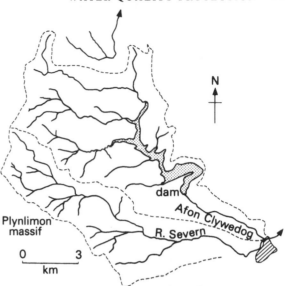

Figure 5.5 Minimum three-day average flows in the River Severn at Bewdley 1921–68. The lowest flow is about five cumecs and the probability of flow falling below eight cumecs is about once every two and a half years. A=prescribed flow at Bewdley before the Clywedog dam construction, B=prescribed flow after the Clywedog dam construction. A *minus* C = additional abstraction upstream of Bewdley allowed by the controlled supply.

Figure 5.6 The catchment of the Clywedog reservoir. Note that the River Clywedog dominates the headwaters of the River Severn catchment.

left very little for industry and domestic consumption, so there was an immediate need for an increase in the dependable supply. The construction of the Clywedog reservoir combined the need for water supply with the bonus of flood control. The main dam, in a steep-sided narrow valley just upstream of the town of Llanidloes, is able to capture the majority of the Severn headwaters (Fig. 5.6). This allows a large amount of water (49 000 000 m³) to be stored behind a relatively small, high dam with minimal loss of land (Fig. 5.7). At the same time, the reservoir protects the towns along the Severn from flooding and enhances the landscape and tourist potential.

For flood control the reservoir level is drawn down by 1 November to leave a capacity of about 8 400 000 m³ (Fig. 5.8). Excess storm water is then allowed to fill the spare capacity by 1 May. From this time on the main function of the reservoir is to augment flow rather than restrain it.

In contrast to the River Severn which experiences short duration floods which can occur at any time of the year, the River Niger has two predictable floods: the Black Flood, which arrives in Jebba about December, and the much larger White Flood, arriving in August. In 1968 a dam was completed to impound a relatively shallow reservoir in the Niger valley upstream of Jebba (Fig. 2.32). Like the Clywedog, this was designed as a multipurpose reservoir including flood control, but in this case the peak flow to be contained was 5000 cumecs (ten times that of a major flood on one of Britain's larger rivers) with a duration of

about a week. When full, the reservoir is 137 km long and up to 24 km wide. For the purpose of flood control it must clearly be at a low level before the arrival of the White Flood and, as with the Clywedog, this is achieved by controlled discharge in the dry season. There must also be sufficient capacity to contain the Black Flood, but this is much smaller. This is achieved by having a maximum capacity in March and a full reservoir in December (Fig. 5.9). While much of the reservoir's operation is for water supply in the dry season, it is very clear that the careful operation of the reservoir has reduced the peak of the White Flood by 60% compared with pre-dam flows and has thus enabled a much fuller use to be made of the potential of the Niger flood plain downstream of the dam.

Water quality protection

Every day about four hundred million cubic metres of water are flushed down the toilets of the United Kingdom ...

Earlier in this chapter we noted that, in general, water supply is taken for granted. But whilst we notice a drought or a flood, we are not made aware of one of the main uses to which water is put – that of effluent disposal. Every day, domestic and industrial consumers use water to remove wastes (**effluents**) out of sight and out of mind. It is one of

Figure 5.7 The Clywedog dam and lake.

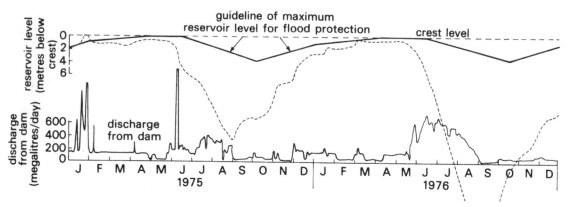

Figure 5.8 Tactical operation of the Clywedog reservoir.

the functions of the water engineer to worry about the outcome. For example, after treatment at an effluent disposal plant (sewage works), the treated effluent – still unfit for use by public water supplies in many cases – is discharged into rivers. It is therefore vital that there is enough water in river networks to allow a satisfactory dilution of this effluent, that which comes straight into the rivers from industrial plant, and in some cases also from storm sewers. This requirement is recognised by act of Parliament under the heading **prescribed** (minimum acceptable) **flow**. At present however

such minimum dilutions cannot be attained because of past neglect, especially on the part of government, in controlling the discharges from factories (Fig. 5.10). Even when suitable treatment is available, there are still times when the treatment works are by-passed. After heavy rain, foul sewers (those carrying untreated effluent) soon fill and untreated overspill passes via the storm sewers directly into the river system (Fig. 5.11).

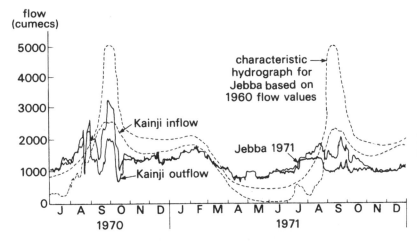

Figure 5.9 Tactical operation of the Kainji reservoir.

Figure 5.10 Effluent from a chemical factory into the tidal reach of the River Tees.

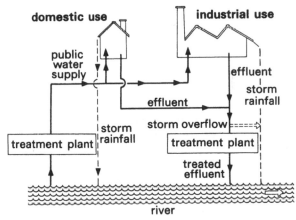

Figure 5.11 The public water supply and its relation to urban hydrology.

In an attempt to get to grips with the problem of water quality, an assessment of the situation was required and to this end a classification scheme was adopted in England and Wales (Table 5.1).

Table 5.1 The quality of natural waters in England and Wales (1973).

	Rivers		Tidal reaches	
Quality class	km	%	km	%
1 (unpolluted)	28 000	78·1	1500	50·7
2 (doubtful)	5000	14·1	640	22·3
3 (poor)	1500	4·2	420	14·6
4 (grossly polluted)	1300	3·6	360	12·4

Clearly, far too many of our rivers are contaminated by effluents to an unacceptable degree, but the fact that it is not worse can be ascribed to the luck that most of our conurbations are on tidal reaches, with only the West Midlands and West Yorkshire some distance inland.

As an example of the problems to be faced with inland industrial locations, consider the case of the River Tame, a tributary of the Trent draining part of the West Midlands conurbation of 2 100 000 people (Fig. 5.12). Industry and domestic users discharge effluent at an average rate of nearly 8·0 cumecs into a river with an average low flow of only 14·0 cumecs so that the dilution of effluent is extremely low. With a natural dry weather flow of 7·6 cumecs there is more effluent in the river than diluting water! In this metal-working area even the natural runoff from streets after rainfall is polluted

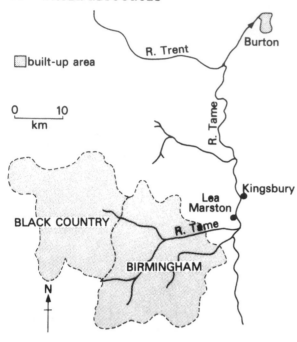

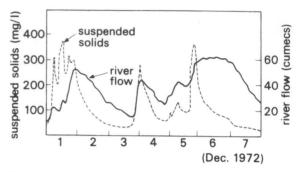

Figure 5.12 The River Tame has a large part of its catchment dominated by the West Midlands conurbation.

Figure 5.13 The problem: the River Tame at Lea Marston showing the relationship between river flow and suspended solids in December, 1972.

because the rainwater picks up pollutants from the atmosphere and the surfaces over which it travels. Contamination is so severe that the water quality of 225 km of the River Trent is affected as well. No fish can survive in the Tame or in the Trent for 16 km downstream of its junction with the Tame. But it is even more incredible to find that the situation

Table 5.2 River Tame water quality at Lea Marston (in milligrams per litre).

	1962–4	1968–70	1975–6
biochemical oxygen demand (BOD)	31	21	20
suspended solids (SS)	150	70	47
ammonical nitrogen (AN)	16	8	5
dissolved oxygen (DO)	3	6	6

Table 5.3 River Tame water quality consequent on a storm, 4 November, 1969.

Time (hr)	Flow (cumecs)	BOD (mg/l)	SS (mg/l)	AN (mg/l)	DO (mg/l)
5·30	4·6	23	120	8	4
6·45	9·3	64	590	9	0
8·45	17·7	93	770	4	2

was once worse (Table 5.2). As is indicated in Table 5.3 and Figure 5.13, the most heavily polluted conditions occur during the first few hours of heavy rainfall when many of the foul sewers are of inadequate capacity and a surge discharge of storm sewage and contaminated water is passed directly to the river without treatment of any kind. However, it is clear that it would be a considerable improvement if the storm surge could be accommodated within a special lake where settling and purification could take place. In fact this is a workable proposition for the Tame, using old gravel pits for river retention (Fig. 5.14). It has been estimated that a retention of five days by a lake system will remove half of the suspended solids and improve the oxygen deficiency by a quarter at relatively low cost.

However, Britain is in many ways fortunate in its water quality management. The main advantage is in having much industry at coastal locations so that only rivers draining through the West Midlands and West Yorkshire are grossly polluted for large stretches. Other countries have more serious problems in that much of their industry and population are located inland. West Germany is a case in point, where, due to the concentration of resources around the Ruhr and the excellent communications afforded by the Rhein catchment, most industry is found a long way inland.

At present 10% of West Germany's land surface is under urban functions, an increase of 50% from 1938. The 61 million people produce an effluent load of $16\,000 \times 10^6$ m³/yr, much of which is sent, either treated or untreated, into the River Rhein. As is common in industrialised countries the rise in pollution has been much faster than population growth; for instance the waste water load carried by rivers and canals rose by 48% in only twelve years prior to 1969. The result is that more than half of the surface-water resources of West Germany are classified as poor or grossly polluted (Fig. 5.15). One of the worst, the River Rhein itself is so contaminated that one cannot identify an object more than 4 cm below its surface at the Netherlands border. But in addition to domestic and industrial sewage, whose characteristics are similar to those described for the River Tame, there is also a large problem from environmental chemicals such as biocides, detergents and fertilisers. Fertilisers, for example, are leached through the soil and taken up in the river water where they cause a rapid growth of the algal population, which explains why the Rhein appears a green colour. Some towns like Stuttgart have to receive clean water by pipeline from over 250 km away because the local rivers are too polluted to be treated economically. In the Ruhr a partial, but local, solution has been to operate the Ruhr dams in the Sauerland to give a constant potable supply of 20 m³/s along the Ruhr at least as far as Essen. At the same time sewer arrangements conduct foul water away to the nearby River Emscher, but the consequence is that this river is little more than an open sewer and needs a treatment plant at its mouth to handle all the flow if any improvement is to be effected in the Rhein.

It is because of this extreme problem that much energy has been expended on finding economical treatment methods. Clearly the Rhein cannot be passed *en masse* through settling tanks or a treatment works, so the maximum use has to be made of **self-purification**. This is the capability of a river to

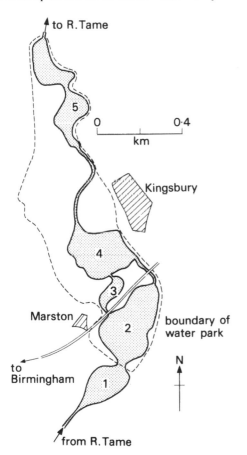

Figure 5.14 The solution: The River Tame lake purification scheme. The primary purification lake (1) has no recreational use but the other lakes (2–5) can form the basis of a water recreation park.

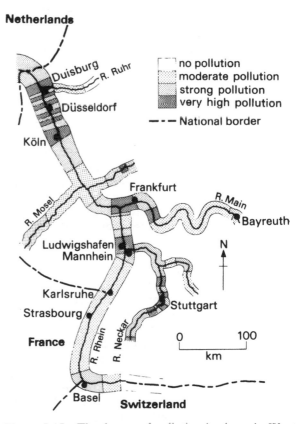

Figure 5.15 The degree of pollution in rivers in West Germany, 1967.

process waste water so that, after a variable distance from the point of introduction, the water body will show approximately the same condition as above the contamination point. River purification is basically a biological process and fundamental to this is the role of bacteria which are able to grow on complex compounds like hydrocarbons, proteins and fats, and break them down into simpler substances such as nitrates and phosphates. The breakdown process uses a great number of bacterial groups, but all are liable to destruction by toxic substances like heavy metals or phenols. It is thus particularly important that some of the most toxic substances are kept out of rivers, for whilst sewage reduces the oxygen content of surface waters (by introducing organic substances which consume oxygen and cause undesirable growth of biological matter in the water), toxic substances can prevent sewage being treated at all.

The effective control of river quality must therefore be a major factor in determining man's need to interfere with the hydrological cycle. With low quality control in some areas, particularly inland, considerable changes in the natural hydrology will be needed on top of those required for the net supply of water.

A strategy for Britain

Because of increasing demands on the water supplies of Britain, it is clear that there will have to be progressively greater interference with the natural hydrological cycle. Each demand has its own requirements – public water supply, effluent treatment, flood control, all need urgent attention. But it is also clear that these demands cannot be treated in isolation, because they are all interdependent: it is not possible to affect the hydrological cycle in one location without repercussions in other places. It is therefore essential that major engineering schemes – for this is what the practical face of hydrology amounts to – are planned in such a way as to make the best total use of the water resource over all the area to be affected. The most sensible operating unit for this sort of planning is the major catchment and this is why the water authorities in Britain are organised around and named after such areas, e.g. Thames Water Authority. But in some areas the demands of one catchment exceed its potential supply and so linking of catchments is necessary, as is the case with the Severn/Trent Water Authority.

The need for massive alterations to the natural flow of Britain's rivers is inescapable. Public demand for water increases at an alarming rate which can only be satisfied by increased storage, even after maximum multiple use by abstractors (Table 5.4).

Table 5.4 Public water demand in England and Wales.

Year	Population	Total demand $m^3 \times 10^6/day$	$l/head/day$	
1900	32·2	3·2	100	
1929	39·5	5·0	127	
1940	41·7	6·0	145	
1961	46·1	10·5	227	
1971	48·6	14·1	290	
Projected				
1981	50·8	17·5	345	} results from one
2001	55·5	24·6	445	} of several models

Table 5.5 Surface- and ground-water quantities abstracted, 1974 (in millions of cubic metres).

	Surface water	Ground water	Total
public water supply	3610	2020	5630
CEGB	5470	—	5470
other industry	2460	500	2960
agriculture	50	40	90
Total	11 590	2560	14 150

Demand by industry other than that obtained via the public water supply is equally voracious (Table 5.5). But there is no absolute shortage of water. The problems basically arise from the fact that most of our rain falls on the Western hills and mountains whilst most people and industry are located in the lowlands. The use of natural resources, like coal, make the geographical pattern of demand for water more unevenly distributed still. Much of the coal for power stations is mined in the Yorkshire, Derby and Nottinghamshire coalfields and transport costs dictate that electricity should be generated nearby. In fact 95% of the CEGB fresh-water cooling load is carried by only four of the principal river systems in Britain: Trent, Mersey, Severn and Yorkshire, and 55% by the Trent alone. The water requirement is such that the gross cooling flow through power stations on the Trent is actually about thirty to forty times the river's dry weather flow. It is simply not going to be possible to alter the river flows in Britain to such an extent that the doubled requirement of the CEGB

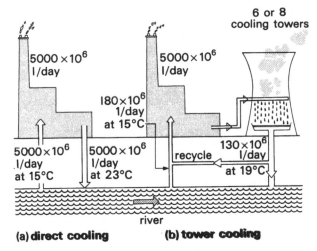

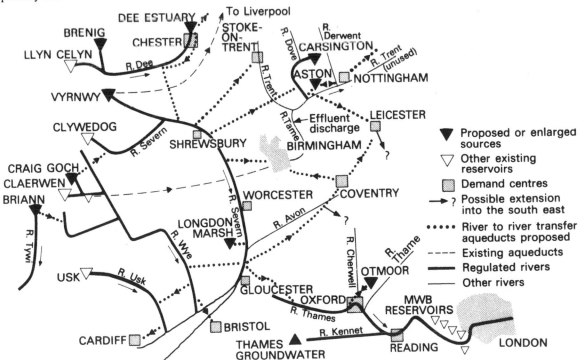

(a) direct cooling **(b) tower cooling**

Figure 5.16 The influence of hydrology on cooling systems: (a) direct cooling requires vast amounts of water from the river which may not always be available; (b) tower cooling loses water by evaporation, but the net loss is relatively small and the gross demand on the river also comparatively small.

Figure 5.17 A possible strategy for water supply in central England and Wales by the year 2001. Note especially the intercatchment transfers.

by the year 2001 can be met, and this is one of the prime reasons why most existing inland capacity has cooling towers beside it (Fig. 5.16) and why new capacity is being built on the coast to employ sea water.

A *national* strategy is needed because supply and demand cut across so many local and regional boundaries. Figure 5.17 gives part of one version of several strategies for the year 2001. Although the Severn and Trent river systems have been closely interlinked for a long time because of the demands for potable water in the Trent catchment, it is envisaged that major exchanges with the rest of the country will also be needed by 2001, especially with the South East.

This sort of need was most dramatically brought home to everybody in Britain in the 1975–6 drought period when rainfall over the sixteen months prior to August 1976 was only 64% of the average. The exact deficit was variable and so were the catchment responses. Catchments on impermeable rock had substantially decreased river flows after only one or two months into the drought. By contrast, rivers (like the Kennet and Ely Ouse) draining permeable areas of chalk only became seriously affected when the drought had lasted for over ten months. In the 1975–6 drought, such low flows did not come until half way through

1976 and so the crisis of supply in the East and South was later than in the West and North. In fact there was never a shortage of water in ground-water storage throughout the drought but, as water tables fell below river beds, the only way to get the water was by pumping and there were just not enough pumping stations to cope. The way in which such resource problems were managed is best illustrated by the following passage:

The North Central Division covers a wide stretch of the Dales ... but the area most vulnerable to drought is that embracing Leeds and Harrogate. Supplies to these centres come from a chain of reservoirs in the Washburn Valley ... As the drought bit deeper the engineers regulated the levels of these reservoirs so that the space to be filled was roughly in proportion to the catchment area of each: this avoided any chance of the lowest reservoir being full when it rained, and losing all its increment in overspill.

But it did not rain. The Leeds reservoirs were only half full in the middle of May. They held that level for a while as a result of additional supplies from the river Wharfe and consumer economies resulting from the publicity campaign, but by mid-July they were down to 41·7% of capacity, and got lower every week: 39·4%; 35·5%; 34·2% in early August. By the end of the month they were below 30%.

North Central Division was better situated than South Western Division, in that some augmentation was possible. The programmed measures to preserve supplies for Leeds and Harrogate for as long as possible ran as follows: Hosepipe restrictions; publicity campaign; reduction of pressures in the distribution system; drought order to permit reductions in compensation flow from reservoirs; drought order to permit temporary abstraction from the River Wharfe, subject to maintaining a prescribed minimum flow; on the assumption that the Wharfe would fall to that prescribed minimum flow, a drought order to permit temporary abstraction from the River Nidd, supported by releases from the Gouthwaite regulating reservoir higher up that river; further reduction of distribution pressures; drought order for provision of supply by standpipes; drought order for wider powers of rationing under the new Drought Act.

The hosepipe ban was placed on Leeds in January 1976 – it had been operating in Harrogate since the previous year. Reduction in reservoir compensations was authorised at the end of March, abstraction from the Wharfe at the end of April, and abstraction from the Nidd by late May. The publicity campaign and the first reduction in distribution pressures were started in April. In April also the Authority had budgeted for the expenditure of £600 000 on the purchase and erection of standpipes alone – not to mention all its other emergency operations throughout the region.... Yorkshire was lucky, or maybe just far-sighted, in that abstractions both from the Wharfe and from the Nidd could be fed into the Leeds supply system without any great expenditure on emergency mainlaying. There were existing trunk mains crossing both rivers. The Wharfe abstraction was pumped into the main that ran from Swinsty Reservoir to Leeds, and the Nidd water entered a trunk main from Leighton to Swinsty.

So, as Yorkshire inflexibly maintains, these operations involved no startling innovations, just thorough planning and hard work. It thought problems out and solved the practical difficulties as they arose. And the Yorkshiremen return again and again to the benefits accruing from reorganisation. Within all divisions they have brought about interconnections and transfers that would have been out of the question before (the Water Act of) 1974. Without that 'switchability', pockets of Yorkshire would have been dry by mid-summer, and local councillors would have been going round jingling their money in a quest for bulk supplies. Without an integrated command and transferable resources of labour and finance there could also never have been the acceleration of the capital programme that took place in the summer of 1976.

Andrews (1977)

The drought forced the general public to recognise the critical nature of water resources. The effects of such a drought can never be catered for on economic grounds, but the long term supply problem does have to be solved. In England and Wales the preferred strategy is: (a) to use as much ground water as possible; (b) to redeploy and enlarge many existing reservoirs; (c) to develop estuary storage at one or two sites; (d) to try to recharge aquifers artificially where possible; (e) to clean up polluted rivers like the Trent so they can be used for potable supplies; and (f) to build a number of new aqueducts to transfer water between catchments. It is clearly a concept of considerable complexity and many of the decisions made are value judgements based on social considerations as well as on economic grounds.

Conclusion

It is clear that additional control over the water resources in Britain will have to be exercised in the years ahead. The exact amount of extra provision needed depends on the accuracy of forecasts of demand and, as demands are prone to change, the water industry cannot plan too far into the future. For example, the demands forecast for 2001 in the late 1960s have recently been revised to only two-thirds of the earlier estimate. Clearly too, there are many conflicts of interest arising mainly because of the large amount of land that reservoirs occupy. There are many arguments for a wide range of alternative supply schemes as exemplified above. It is all the more important, therefore, that the planning of water resources be based on a sound understanding of the hydrological cycle (Ch. 2) and the way it affects particular catchments (Ch. 3). With such understanding it should be possible to make the most advantageous use of the limited land available to us.

Example problems

A. Figure 5.18 shows the natural flow of the River Thames between 1974 and 1977. (a) Estimate the point at which the Water Authority would have become worried about providing enough water in the river. (b) Describe some of the steps you think might have been taken to alleviate the water supply problem. Relate this to newspaper articles of the 1975–6 drought and to *'We didn't wait for the rain'* (see Further Reading) to see what was really done.

B. Reservoir construction is one of the topmost priorities in applied hydrology, but the success of site proposals brings hydrology into conflict with other environmental considerations. Three reservoir schemes that were proposed in the 1960s are outlined below. Only the Kielder scheme is nearing completion. Using the first example for reference, discuss the range of factors that would influence planning permission for the other examples. Examine the site in an atlas and, if possible, on a 1:25 000 OS map. The appropriate map is referred to in each case. Also make reference to Figure 5.17.
a. A new reservoir proposed in the Kielder Forest area of Northumberland, Kielder water (now under construction) (grid area NY 7087).

This reservoir would produce a daily yield of 955×10^3 m³ which requires a storage of 200 000 m³ of water inside the Kielder Forest. The land is mainly of low grade, with conifer and permanent pasture, mostly owned by the Forestry Commission, but some of the 11 km² of land required was in private ownership. The major landowner was, in this case, in favour of the proposed reservoir which simplified land acquisition. The Countryside Commission and Nature Conservancy considered that the reservoir would make an attractive landscape feature with considerable recreation potential. It is also large enough to reduce the need for other reservoirs in the area, some of which would be in areas of higher agricultural potential or in National Parks. The reservoir would be 50 m deep at the dam and the considerable amount of storage involved could be used to regulate the upper Tyne for flood control while allowing transfers of water to the Wear and the Tees and eventually to the Yorkshire Ouse system.
b. A proposal for a new reservoir near Great Bradley, Cambridge/Suffolk border (grid area TL 6754).

The area of land occupied would be the same as Kielder water. (i) Why should the storage capacity be 99×10^3 m³ and why would its yield be restricted to 215×10^3 m³/day? (ii) Is there a large or small demand for water in this area? (iii) Could the reservoir be usefully put to amenity and landscape use in this area? (iv) What quality of farmland would be expected in this area and would the flooding of three large farms, twenty-two dwellings and a nature conservation site be a major handicap? (Note, other reservoirs like this have been built, for example Graffam Water near Northampton.) (v) On balance would this scheme be likely to be rejected, accepted or shelved?

C. A proposal to use the Dee estuary for four bunded reservoirs (grid area SJ27/37).

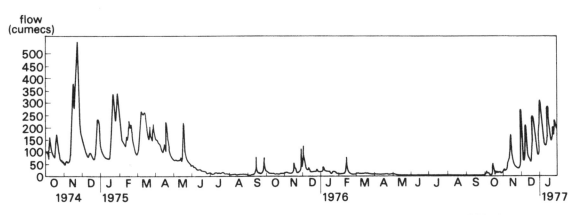

Figure 5.18 Average daily flows of the River Thames at Teddington, 1974–6.

This site would be in the upper Dee estuary, at present salt marsh. (i) Is this a large or small demand area? (ii) Could the reservoir be useful as an amenity to local urban areas? (iii) What effect would embankments have on the local landscape and would this be acceptable? (iv) What impact might there be on the natural fauna and flora? (v) Would the improved road access to the area be of long-term benefit? (vi) On balance would the scheme be likely to be accepted, rejected, or shelved?

D. Hydro-electric power is of only minor importance to the total power requirements of Britain, but plays a major role in the economies of some other countries. Discuss the hydrological factors involved in decision making for the provision of (a) HEP in a mountainous region such as Norway and (b) a lowland region such as the Rhine Rift valley.

GLOSSARY

aquifer A rock having sufficient voids within it to be able to conduct significant quantities of water.

base flow The low flow in a river due to ground water and slow throughflow, which occurs when there has been no rainfall for a prolonged period.

capillary action The process by which moisture moves in any direction through the smaller pores and in films around particles.

capillary moisture The soil water held in small pores and in films around particles after the soil has drained under gravity.

catchment The area of land whose surfaces focus water towards a river system upstream of the point in question.

cohesion The ability of a soil to be held together by surface-tension water films.

contributing area The part of a catchment, usually adjacent to the stream, that contributes most of the water to the hydrograph peak.

cumec The common unit of river flow measurement (=cubic metres per second).

effluent Waste products carried by water in suspension or solution.

eluviation The removal of soil materials in suspension.

evaporation The loss of water from a surface as vapour.

evapotranspiration The combined result of **evaporation** and **transpiration**.

exponential A term used to denote a characteristic shape of curve having an initially rapid decline but becoming progressively slower with time.

factor of safety (F) The ratio of **shear stress** to **shear strength**.

field capacity The moisture content prevailing in a soil after all surplus water has drained away under gravitational influences and before there has been significant loss by **evapotranspiration**.

flush A boggy region of a hillside which often contains subsurface and surface channels for water flow.

gleying A process occurring in a soil which is due to waterlogging and involves a colour change to grey or blue tones.

hydraulic gradient The rate of change in **hydraulic head** with distance.

hydraulic head The pressure applied due to the weight of water above a given point.

hydrograph A plot of river flow against time.

hydrological cycle The continuous movement of water over the Earth's surface which involves **evaporation** from, and eventual return of water to, the oceans.

infiltration The passage of water across the soil surface and into the soil.

interception The trapping and retention of **precipitation** before it reaches the soil or river channels.

leaching The removal of soil materials in solution.

overburden pressure The pressure exerted on a point resulting from the weight of soil or rock above it.

overland flow The flow of water over the soil surface.

percolation The movement of water in a soil in a vertical direction.

permeability The property of a soil or rock which allows it to conduct water.

pipeflow Movement of water through soil via connected voids of a size larger than normal which in section often resemble small pipes.

pore pressure The force on water in a soil pore due to an imposed **hydraulic head**.

porosity The proportion of void space in a soil.

precipitation A group term for all forms of water deposition from the air. Although usually applied to rainfall and snowfall, it can also refer to dew, rime, etc.

recession The decline in river flow after the storm contributions have passed.

runoff The output of water from a **catchment** or part of a catchment.

shear strength The force that resists a **shear stress**.

shear stress The force applied to a body encouraging it to slide.

soil creep The slow downslope movement of soil material due to expansion and contraction of the clay fraction with wetting and drying cycles.

throughflow The movement of water in a soil in a downslope direction.

transpiration The loss of water from the surface of a living substance as part of its respiratory process.

unconfined aquifer An aquifer which is connected directly to the ground surface by permeable strata over all of its upper surface.

water table The upper surface of the saturated zone in soil or rock.

weathering The *in situ* decomposition of material.

wilting point A soil moisture content so low that water remaining in the soil is held too tightly to be available to plants.

FURTHER
READING

Most books on hydrological topics are at a somewhat advanced level and are suitable for reference only. However:

Water by L. Leopold (Freeman, 1974) is an elementary and very readable text with an American slant.

Water in Britain by K. Smith (Macmillan, 1972) deals with water resources and has much detailed information.

Principles of hydrology by R. C. Ward (McGraw Hill, 2nd edn, 1975) is a standard treatment of all aspects of scientific hydrology, but is rather academically biased.

Drainage basin form and process by K. J. Gregory and D. E. Walling (Arnold, 1973) is an excellent, but again academic, book dealing not just with hydrology but also with its overlap with fluvial geomorphology, and is an especially good adjunct to Chapter 4.

Water, earth and man edited by R. J. Chorley (Methuen, 1969) is a comprehensive collection of scientific papers dealing with all aspects of water.

Parts of *Conservation in practice* edited by A. Warren and F. Goldsmith (Wiley, 1974) deal with aspects of hydrology and soil study in a broader context.

Probably the most easily read and accessible source of current hydrological topics are articles in the *Geographical Magazine*, some of which are listed below.

    The rain in Spain (April 1974, 338–43)

    Water across the American continent (June 1974, 472–9)

    Deluge in Australia (June 1974, 465–71)

    River in the frozen north (August 1974, 634–40)

    Mighty rivers show their strength (November 1971, 131–5)

    Cavernous depths of Yorkshire (October 1972, 36–43)

    New role for Mother Ganges (June 1973, 666–8)

    Man made oases of Libya (November 1972, 112–15)

    Snowy mountains bonanza (April 1977, 424–9)

    An imbalanced resource (May 1977, 489–92)

    Money down the drain (May 1977, 493–7)

    Give us this day . . . (May 1977, 498–501)

    Suburbanisation with salt (May 1977, 502–4)

    Conservation in Kano (May 1977, 504–7)

    Island (Malta) with few options (May 1977, 508–9)

The most readable book on the 1975–6 drought is '*We didn't wait for the rain*' by C. D. Andrews (National Water Council, 1977).

The Soil Survey of England and Wales (Harpenden) and the Soil Survey of Scotland (HMSO) publish regional *Soil Memoirs* which contain a wealth of data and a full colour map of each area. They also publish a series called 'Soil Record' which is of a more general nature. A list of these can be obtained from the Soil Survey of England and Wales, Rothamstead Experimental Station, Harpenden, Herts, and The Macaulay Institute for Soil Research, Aberdeen AB9 2QJ respectively.

There are several books dealing with basic processes in soil study and geomorphology. Amongst these, *Landscape processes* by D. and V. Weyman (Allen & Unwin, 1977) adopts an approach towards landscape evolution similar to that of Chapter 4. *Fluvial processes in geomorphology* by L. Leopold, M. Wolman and J. Miller (Freeman, 1964) is a very comprehensive though somewhat technical treatment. *Soil processes*, B. J. Knapp (Allen & Unwin, 1979) deals more fully with the processes of soil formation needed to underpin aspects of Chapter 3 and contains many more specific examples.

INDEX

9 780045 510306